FLIGHT

for the Private Pilot

Book 3

Aviation Book Series

Dr Stephen Walmsley

Disclaimer
Whilst every effort has been made to ensure the accuracy of the information, the author does not give any guarantee as to its accuracy or otherwise. Nothing in the contents of this book is to be interpreted as constituting instruction or advise relating to practical flying. Students preparing for their relevant exam should consult training their syllabus published by the relevant authority to ensure they are studying towards the most up to date syllabus. The author shall not be liable nor responsible to any person or entity concerning errors and omission, or loss or damage caused directly or indirectly by the use of the information contained in this book.

ISBN: 9798454208493

AVIATION BOOKS SERIES

Flight Radio for the Private Pilot is part of the Aviation Books Series that provides the reader with an educational and enjoyable reading experience. A focus has been placed on practical, hands-on aviation by linking science with the real world. Books in the series include:

Book 1: Human Factors

Book 2: Aviation Weather

Book 3: Flight Radio

Book 4: Principles of Flight

Book 5: Flight Navigation

CONTENTS

FLIGHT RADIO

Copyright

Aviation Books Series

Preface

Introduction   1

Chapter 1: Radio Waves   4

Chapter 2: Aeronautical Frequency   11

Chapter 3: The basics   17

Chapter 4: Basic phrases   25

Chapter 5: Call signs   32

Chapter 6: Basic flight radio techniques   44

Chapter 7: Departure   56

Chapter 8: Cross-country   66

Chapter 9: Arrival   83

Chapter 10: IFR radio calls   94

Chapter 11: Radio failure   101

Chapter 12: Emergency   110

Chapter 13: Radio Equipment   121

Conclusion   127

Index   129

Books in this series   131

PREFACE

The aircraft radio is an amazing piece of equipment. By pressing down on the microphone switch you can have a conversation with a wide range of people; from the local controller to other pilots in the area. But operating the radio also comes with responsibility. To avoid the serious consequences that can result from miscommunication, it is essential that all pilots have a good understanding of flight radio.

This book follows closely the syllabi of Flight Radio from a range of aviation authorities around the world, allowing the reader to obtain the required knowledge in Flight Radio. This book goes beyond these syllabi, with a particular focus on practical aviation, linking science with the real world. Each chapter contains a range of visual figures in full color and mini case studies that will allow the reader to have a deeper understanding of the wide range of components of flight radio.

INTRODUCTION

Imagine you have completed your pre-flight checks and the weather looks great. All you need to do now is push down on the microphone button to get your flight moving. When learning to fly, this simple task can seem like one of the most stressful events. What happens if I forget what to say? What if I say the wrong word? What if air traffic control tells me off? You don't need to be a new flyer to have these concerns. Pilots that have taken a break from flying or generally remain clear of controlled airspace may also feel a little nervous operating the radio. Fortunately, like many of the skills you learn, operating the radio comes easier with practice. In no time at all, speaking on the radio will become second nature. But to master this important skill, you also require the underlying knowledge.

Flight radio can sound like a foreign language or a secret code. However, history has repeatedly shown that a communication breakdown can lead to misunderstandings, or worse, a major disaster. There are many barriers to effective communication in flight, including noise, vibrations, radio clutter, and distractions. There are no traffic lights in the sky to let pilots know if they can safely continue or not. Effective communication between pilots and air traffic control is one of the key tools used to ensure aircraft remain safely separated from each other, especially in and around busy aerodromes. At busy aerodromes, an aircraft could be taking off and landing every minute, and without clear and effective communication, all sorts of problems would result.

This book explores a wide range of flight radio components that are essential for pilots to master to operate an aircraft safely. You will begin with the most basic component of radio science – the radio wave. To many people, radio science is a mysterious subject, but you will see a basic understanding is helpful to appreciate some of the flight radio limitations. You will then take a look at the building blocks of flight radio, which is based on how letters and numbers are spoken. You will then take a flight and explore a range of typical radio calls you may be required to make, from taxi to landing, at towered and non-towered aerodromes. In the final chapters, you will look at how the flight radio can be an essential tool when things go wrong, such as in an emergency.

Although some of the communication practices that are covered in this book may sound strange, they are designed to provide a layer of protection to avoid miscommunication. When you are flying on a nice sunny day, with a low workload, you are unlikely to have any trouble communicating clearly. But what happens if you are under a little stress; the weather is poorer than expected, there is time pressure to complete the flight and the radio is playing up. Did the controller just ask you to descend to nine thousand feet or was it five thousand feet? Did you have to taxi to holding point 'C' (see) or was it 'D' (dee)? When a pilot is strained and stretched to their limits, the common flight radio practices that you will explore in this book provide a vital tool to avoid confusion.

CHAPTER 1: **RADIO WAVES**

PILOT: *Tower, verify you want me to taxi in front of the 747.*
TOWER: *Yeah, it's OK. He's not hungry.*

T he aircraft radio is an amazing piece of equipment. By pressing down on the microphone switch you can have a conversation with air traffic control, who might be just a few hundred feet away in the aerodrome tower or hundreds of miles away in a control center. Although the main focus of this book is how to speak on the radio, it is just as important to have a basic understanding of how the radio works. This basic knowledge will help you understand why you are having trouble contacting air traffic control and what you can do to fix the problem.

Radio Waves

Radio waves are the key component that allows you to speak on the radio and listen to other radio users.

Radio waves can be a tricky topic as it can be difficult to visualize something that we cannot see. It is easier to think of radio waves like waves in water. Imagine dropping a large stone in a completely still lake. As the stone splashes into the water, a wave radiates out. If you looked carefully at the wave that forms, you would notice it has a series of peaks and troughs.

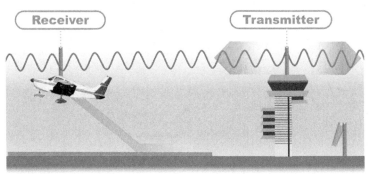

Figure 1.1: *A transmitter is required to create a radio wave, which can be detected by the antenna on aircraft in the area.*

A radio wave is similar to the wave formed in the water; however, a radio wave does not need a medium to travel through (e.g. water or air) and is partly electric and partly magnetic, which is known as an electromagnetic wave. A radio wave is basically a packet of energy that can be used to help talk or listen on the radio. In fact, the radio wave is just a small part of a wide range of electromagnetic waves, which include X-rays. To create a radio wave, a transmitter is required, along with an antenna that sends the radio wave out, as shown in Figure 1.1. Likewise, to detect a radio wave, an antenna is also required. Antennas are normally designed to only pick up a limited range of frequencies, for example, the antenna on the aircraft can only detect frequencies within a narrow band of aviation frequencies (which we will discuss in more detail in the next chapter). As the radio wave travels further from the transmitter, it tends to weaken – just like the wave made in a lake that gets smaller and smaller. As a result, the receiver normally also requires an amplifier, which can enhance the radio wave.

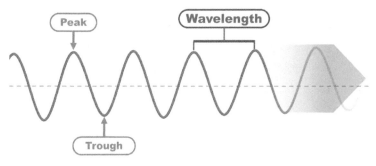

Figure 1.2: *A radio wave has peaks and troughs. If you measured the length of one full cycle, you would find the wavelength.*

If you took a closer look at the radio wave as it went past you, you would see it contains some very important characteristics. Firstly, it would pass you very quickly, in fact, radio waves travel at the speed of light, which is a staggering 300,000,000 meters per second. Also, recall the radio wave is moving past you in a wave shape – peaks and troughs. One full wave is known as a *cycle*, and if you measured the length of one full cycle you would find the *wavelength*, as shown in Figure 1.2. A radio wave can have a wavelength from around 1 millimeter to over 100 kilometers.

The final, and most important characteristic is *frequency*. Frequency is the number of cycles that occur in one second and is measured in hertz (Hz). If one cycle passed you in one second, the frequency would be 1 Hz, two cycles would be 2 Hz. Most radio waves have a considerable number of cycles per second, as a result, the number of hertz is normally abbreviated. One thousand hertz is known as a kilohertz (kHz), which means 1,000 cycles passed per second. One million hertz is known as a megahertz (MHz), and one billion is known as a gigahertz (GHz).

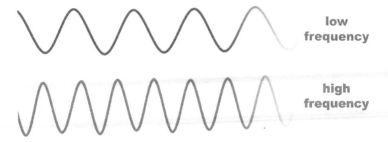

low frequency

high frequency

Figure 1.3: *Lower frequencies have a longer wavelength, whereas higher frequencies have a shorter wavelength.*

Frequency and wavelength have an indirect relationship. The higher the frequency, the shorter the wavelength, as shown in Figure 1.3. We can look at this simple relationship by comparing two frequencies. A radio wave with a frequency of 1 Hz would result in one cycle per second. Remember a radio wave travels at the speed of light (300,000,000 meters per second), therefore the wavelength would be 300,000,000 meters. If the frequency was 2 Hz, this would result in 2 cycles per second, therefore the wavelength would be 150,000,000 meters (half the wavelength compared to 1 Hz). Practically, you will not be pulling out your calculator to work out the wavelength of a frequency during a flight, but it is worth noting, the higher the frequency, the shorter the wavelength, and vice versa.

Radio Spectrum

Radio waves are split up into several sub-groups depending on their frequency, which makes up the radio spectrum. The radio spectrum starts at less than 30 kHz (very low frequency) and goes up to 300 GHz (extremely high frequency).

Frequency band	Frequency
Very Low Frequency (VLF)	< 30 kHz
Low Frequency (LF)	30 – 300 kHz
Medium Frequency (MF)	300kHz–3MHz
High Frequency (HF)	3 – 30 MHz
Very High Frequency (VHF)	30 – 300 MHz
Ultra High Frequency (UHF)	300MHz –3GHz
Super High Frequency (SHF)	3 – 30 GHz
Extremely High Frequency (EHF)	30 – 300 GHz

Different parts of the radio spectrum are utilized by various users, as shown in Figure 1.4. For example, frequencies towards the lower end of the scale are more suitable for submarine communication. If you are tuning up an AM radio station in your car, you are around the medium frequency band. Satellite communication tends to utilize higher frequencies.

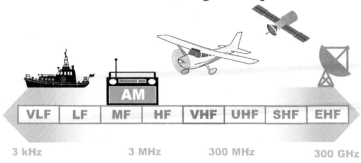

Figure 1.4: *Different parts of the radio spectrum are utilized by various users.*

Each frequency band offers different advantages and disadvantages, especially in regards to range and transmission quality. You may be wondering where do aviation frequencies sit within this radio spectrum? In the next chapter, you will see aviation radio frequencies are all squeezed into a tiny part of the radio spectrum.

CHAPTER 2: **AERONAUTICAL FREQUENCY**

PILOT: Tower, give me a rough time-check!
TOWER: It's Tuesday, Sir.

When you peer down at an aerodrome chart and spot the frequency you require, in most cases, it will not indicate where on the radio spectrum it belongs. For example, a Tower frequency of 118.3 or an ATIS frequency of 124.1, as shown in Figure 2.1.

HAMILTON

TOWER		GROUND		ATIS
118.3	124.8	119.5	123.2	124.1

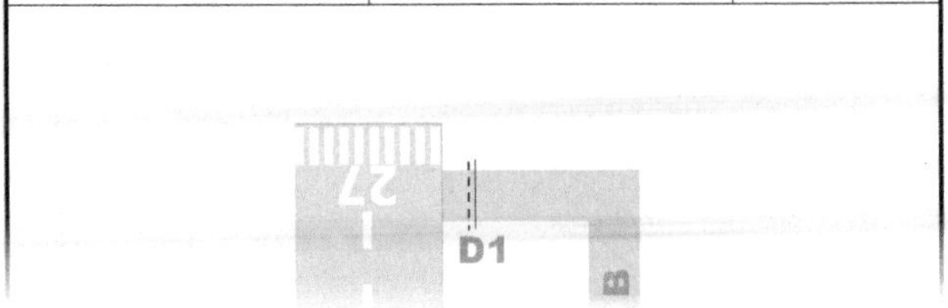

Figure 2.1: *An example of a section of an aerodrome chart with a range of frequencies.*

Practically, this is not an issue, as you simply input your required frequency into the radio unit. However, having an understanding of the part of the radio spectrum you will be operating on will be beneficial. In this chapter, you will see aviation radio frequencies are squeezed into a tiny part of the overall radio spectrum. This narrow range can cause frequency congestion.

VHF Communication

Imagine you have just selected the Tower frequency 118.3 on your flight radio. Where does this frequency sit within the radio spectrum? This frequency is 118.3 MHz, which means the radio wave you are about to create will repeat itself 118,300,000 times every second (118.3 million times per second). If you glanced back to the previous chapter, you will note this sits nicely within the very high frequency (VHF) band. Most aviation radio communication will be within the VHF band. Like all frequency bands, there are advantages and disadvantages of operating within this frequency range. The main advantage is it allows reasonably good quality communication, with limited interference (sometimes called static). Clearly, this is a major plus in aviation, as the overall aim is to avoid miscommunication. Another advantage with VHF frequencies is the antennas are relatively small, which means they easily fit onto aircraft of all sizes. The main disadvantage is range, that is, how far the radio wave can travel. The range of a VHF radio wave is line-of-sight, which means radio reception can be lost if an aircraft travels too far from the transmitter. If you conduct a flight over a large stretch of water you may have trouble communicating with a VHF frequency, as you may find you have gone beyond the line-of-sight of the nearest radio station. A VHF radio wave also has limited ability to bend around obstacles, such as mountains, as shown in Figure 2.2, creating a radio shadow. This may be experienced when flying in hilly or mountainous areas. You will see towards the end of this chapter, high frequency radio and satellite communication can overcome some of these issues.

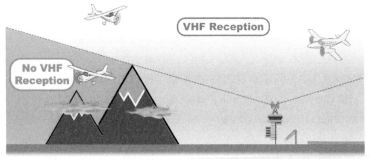

Figure 2.2: *VHF reception is limited to 'line-of-sight'.*

If you looked at the frequencies from a number of different aerodromes, you would find they do not stray too far. This is because aviation has only been allocated frequencies between 118 to 136.975 MHz for flight radio (sometimes called the Airband). The band just below, from 108 to 117.95 MHz is also allocated to aviation and used for radio navigation. For example, if you looked at an ILS or VOR frequency, you should find these frequencies sit within this lower range.

Even a modest-sized aerodrome can have a considerable number of frequencies, such as a Tower, Ground, and ATIS frequency, as shown in Figure 2.1. Around some large cities, it is common to have a number of aerodromes – from large international airports serving commercial airliners to small aerodromes used for general aviation. Each of these aerodromes will have a range of frequencies, which cannot be too similar to surrounding stations. On top of all these aerodromes, frequencies will be required for sections of upper airspace (e.g. Approach, Departure, Center). As VHF range is line-of-sight, aerodromes a considerable distance from each other can share the same frequency. But within the same general area, a large number of different frequencies are required, resulting in the Airband very quickly becoming congested.

Channel Spacing

To squeeze in as many stations as possible, the spacing between them has been kept to a minimum. A station may also be referred to as a channel (e.g. a Tower frequency on 118.3 MHz can also be referred to as a channel). Originally the minimum spacing between channels was 50 kHz. If a channel had a frequency of 120.0 MHz, the next nearest channel would have to be 120.05 MHz, then 120.1 MHz, etc. This provided a total of 360 channels within the Airband. This spacing quickly became inadequate and in most parts of the world 25 kHz is the standard spacing, allowing 720 channels (e.g. 120.0 MHz, then 120.025 MHz, 120.05 MHz, etc). In some parts of the world, especially in Europe, even 25 kHz spacing does not provide enough channels and 8.33 kHz spacing has been introduced in these frequency congested areas, providing 2,280 channels.

Although the VHF radio is the primary means of communicating during a flight, it is not the only one. In areas with limited VHF coverage, such as over the ocean, a high frequency (HF) radio may be utilized. The main advantage of HF is the majority of transmissions are able to travel beyond the line-of-sight and can sometimes reach areas in a VHF radio shadow (e.g. behind a mountain). However, HF transmissions are more prone to interference (signals are not as clear as a VHF transmission). The HF aviation frequency band is between 2,850 to 22,000 kHz. Aircraft require a separate radio unit to transmit on a HF radio, which is uncommon in smaller aircraft (as they do not normally wander long distances out to sea!). Technology is starting to provide another communication solution for flights outside of VHF reception, in the form of satellite communication. Satellite based communication is more likely to be found in larger aircraft, helping to overcome some of the quality issues of HF transmissions. Radio equipment will be explored in more detail in chapter 13, including emerging technologies.

Figure 2.3: *Aircraft flying in areas with limited VHF coverage may need to use an HF radio.*

You now have the Tower frequency tuned up correctly on your flight radio and are ready to make your first transmission. But before you push down on the microphone button to clog the airways, you first need to know how to speak on the radio. Common radio terminology helps reduce the risk of miscommunication and ensures efficient use of the airways.

CHAPTER 3: **THE BASICS**

TOWER: CessnaABC, What are your intentions?
PILOT: To get my Commercial Pilot's License, and Instrument Rating
TOWER: I meant in the next five minutes not years

In 1989, a Boeing 747 freighter was conducting an approach into Kuala Lumpur, Malaysia. Air traffic control radioed to the flight, 'Tiger 66, descend two four zero zero' [2,400 feet]. The captain of the flight heard 'descend to four zero zero' and replied, 'okay, four zero zero' (meaning 400 feet). This simple miscommunication was one of the contributing factors that led to the aircraft descending too low and crashing into a hillside that was 437 feet above sea level. In a similar case, a Boeing 747 was on approach to Nairobi, Kenya, and was cleared to 'descend seven five zero zero feet' (7,500 feet). Both pilots believed they had heard 'five zero zero zero feet' (5,000 feet). Fortunately, the aircraft broke out of cloud in time for the crew to see the terrain and climb back up to a safe altitude.

Some of the communication practices in this chapter

will seem a little strange, but as these cases high-light, a minor miscommunication can have serious consequences. This chapter explores the foundations of flight radio, which is based around how letters and numbers are spoken. The key to remember, the goal is to avoid any possibility of confusion. For example, the number nine should be pronounced *'niner'*. You are probably thinking if you were cleared to *'nine thousand feet'* you would likely understand this instruction correctly. But what if you are under a little stress as the weather is a little worse than expected, you feel a little tired because you did not sleep too well, and your radio is playing up, so it is not as crisp and clear as normal. Did air traffic control just instruct you to climb to nine thousand feet or was it five thousand feet? Some words and letters can sound similar, especially when it is difficult to hear clearly. For example, the letter 'A' could be confused with the number '8', or the letter 'C' (see) sounds a lot like 'D' (dee) or 'V' (vee). To help overcome these problems, a standard way of saying letters and numbers has been devised.

Transmission of Letters

Transmission of letters is the starting point of flight radio, which heavily utilizes the phonetic alphabet. As shown in the table that follows, each letter has a word associated with it, and each word is spoken with certain parts emphasized. For example, A for

Alpha, which is spoken AL fah, with the syllables stressed that are in capital letters.

Letter = Word (Spoken)	Letter = Word (Spoken)
A = **Alpha** (ALfah)	N = **November** (noVEMber)
B = **Bravo** (BRAHvoh)	O = **Oscar** (OSScar)
C = **Charlie** (CHARlee)	P = **Papa** (pahPAH)
D = **Delta** (DELLtah)	Q = **Quebec** (kehBECK)
E = **Echo** (ECKoh)	R = **Romeo** (ROWme-oh)
F = **Foxtrot** (FOKStrot)	S = **Sierra** (seeAIRrah)
G = **Golf** (golf)	T = **Tango** (TANGgo)
H = **Hotel** (hotTEL)	U = **Uniform** (YOUnee-form)
I = **India** (INdee-ah)	V = **Victor** (VIKtah)
J = **Juliet** (JEWleeETT)	W = **Whiskey** (WISSkey)
K = **Kilo** (KEYloh)	X = **X-ray** (ECKSray)
L = **Lima** (LEEmah)	Y = **Yankee** (YANGkey)
M = **Mike** (mike)	Z = **Zulu** (ZOOloo)

Practically, you will use the phonetic alphabet reasonably often. For example, taxiways and holding points often have a letter assigned to them, just like a street name. Such as taxiway B or holding point C. When air traffic control clears you to taxi, they will use the phonetic alphabet (e.g. taxiway *bravo*, holding point *charlie*). Most small aircraft call signs will include letters and therefore almost every radio transmission will contain the phonetic alphabet (e.g. cleared to land, *alpha bravo charlie*). You will explore call signs in more detail in chapter five. There are a few exceptions when letters do not need to be spoken with the phonetic alphabet, with the letters simply spoken like normal. These are normally abbreviations in which there is no risk of confusion. Examples include ILS (instrument landing system), RVR (runway visual range), and QNH (altimeter pressure setting).

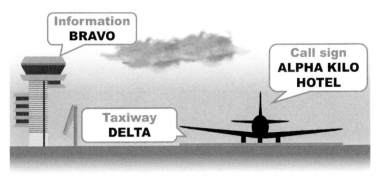

Figure 3.1: *Examples of the phonetic alphabet usage.*

Transmission of Numbers

Like letters, numbers also have specific pronunciations. However, numbers also have an added challenge as to when they should be spoken individually (e.g. *two four zero zero*) or when the words hundreds or thousands should be used (e.g. *two thousand, four hundred*). The following table shows how numbers should be spoken, with the syllables stressed that are in capital letters:

Number / Spoken
0 = ZE-RO
1 = WUN
2 = TOO
3 = TREE
4 = FOW-er
5 = FIFE
6 = SIX
7 = SEV-en
8 = AIT
9 = NIN-er

All numbers, except for a few important exceptions are spoken as individual digits. Below are some important examples:

Number type	Spoken
Heading 320	TREE TOO ZE-RO
Wind 250 degrees at 15 knots	TOO FIFE ZE-RO degrees WUN FIFE knots
Transponder code 1200	WUN TOO ZE-RO ZE-RO
Runway 18	Runway WUN AIT
Altimeter 1020hPa	WUN ZE-RO TOO ZE-RO
Altimeter 29.98inHg	TOO NINer NINer AIT
Flight level FL150	WUN FIFE ZE-RO
Call sign Havocair123	Havocair WUN TOO TREE

Radio transmissions that contain altitudes, cloud heights, visibility, and runway visual range (RVR) that contain whole hundreds or thousands should include the words HUN-dred (hundred) or TOU-SAND (thousand). For example, 2,400 feet should NOT be spoken TOO FOW-er ZE-RO ZE-RO, but rather, TOO TOU-SAND, FOW-er HUN-dred. Other examples include:

Number Type	Spoken
Altitude: 9,500	NIN-er TOU-SAND, FIFE HUN-dred
Altitude 12,000	WUN TOO TOU-SAND
Cloud height: 800	AIT HUND-red
Visibility: 1000	WUN TOU-SAND

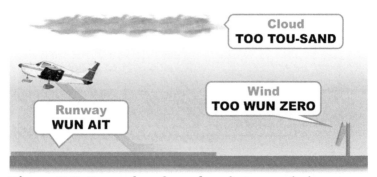

Figure 3.2: *Examples of numbers in transmissions*

Time

Time is transmitted as individual digits, such as the time 09:20 would be *ZE-RO NINer TOO ZE-RO*. However, there are a few other important time-related techniques to note. In aviation the 24-hour clock is used (e.g. 05:20pm = 17:20) and Coordinated Universal Time (UTC) is the standard time zone. Regardless of where you are flying in the world, this standard time zone is used by everyone, reducing the risk of time-related confusion. The word *'zulu'* is often used to emphasize the time is UTC. Local time is sometimes utilized, in which case, the word *'local'* should be transmitted at the end of the time. Usually, only the minutes of the hour are required to be transmitted, for example, TOO ZE-RO for the time 09:20. However, if there is any possibility of confusion, the full time should be provided. Time examples include:

Time	Spoken
10:15	WUN FIFE or WUN ZE-RO WUN FIFE
17:20	TOO ZE-RO or WUN SEVen TOO ZE-RO
05:00	ZE-RO FIFE ZERO ZERO
09:20 local	ZE-RO NINer TOO ZE-RO local

Frequency

Frequency transmissions (e.g. 125.2 MHz) are the final number related radio calls to consider. Frequency transmissions are common during a flight. It is important the correct frequency is communicated clearly. This is where we come across our first area in which different parts of the world use different communication principles. Such differences are normal and generally minor, and in most cases unlikely to cause confusion. Regardless of where you fly in the world, all digits in a frequency need to be transmitted. The decimal point should also be spoken, but the phrase used varies. In the United States, the word POINT is utilized (e.g. 125.2 is spoken WUN TOO FIFE **POINT** TOO), whereas in Europe, New Zealand, Australia, and Canada the word DECIMAL is used, spoken DAY-SEE-MAL (e.g. 125.2 spoken as WUN TWO FIFE **DAY-SEE-MAL** TOO). If a frequency ends in consecutive zeros after the decimal point, some of these may be omitted (e.g. 121.000). In this example, only the first zero after the decimal point is spoken (e.g. WUN TOO WUN POINT ZERO).

You now have a good grasp of the foundations of speaking on the radio, but radio calls contain much more than letters and numbers. If air traffic control asks are you ready to taxi, should you reply, *YES, OK, or ROGER*? You will see in the next chapter, there is a range of words and phrases that are utilized to ensure everyone is answering this question the same way.

CHAPTER 4: **BASIC PHRASES**

TOWER: CessnaABC, turn right now and report your heading.
PILOT: Wilco, 341, 342, 343, 344, 345...

Aviation has adopted a number of standard words and phrases. In the case study that you will explore shortly, you will see there can be serious consequences when non-standard phrases are used in radio transmissions. But there is another important reason to use these standard words and phrases – to reduce radio clutter. In busy airspace, there can be a large number of aircraft using the same frequency. To ensure aircraft and air traffic control can communicate safely, it is important radio calls are not only clear and unambiguous, but also as short as possible. These standard words reduce the risk of miscommunication and at the same time keep radio calls as short as possible.

On March 27 1977 a terrorist incident at the main airport on the Canary Islands, Gran Canaria, resulted in a considerable number of aircraft diverting to the nearby Tenerife airport. At the time Tenerife airport was relatively small, with a small airport terminal, and limited parking space for aircraft. The airport was jammed packed with large aircraft, including a KLM and Pan Am Boeing 747. Once the all-clear was given, the aircraft began to prepare to make the short hop to the main airport. Unfortunately, about the same time fog rolled in, severely limiting visibility. As the taxiways were being used for aircraft parking space, each aircraft needed to taxi down the main runway to ensure they were taking off in the right direction (back-track). To get the aircraft away as quickly as possible, the KLM and Pan Am aircraft taxied onto the runway at the same time. The intention was the KLM aircraft would go to the end of the runway and wait for the runway to be clear before taking off, whereas the Pan Am aircraft would taxi most of the way down the runway before jumping off the runway to a holding point to wait their turn. Due to the thick fog, the pilots of both aircraft could not see each other, nor could air traffic control. Therefore, communication was the only way of making sure everyone was doing what they should be.

The KLM aircraft had lined up and was ready to go. They were under time pressure to complete the flight, but with the thick fog, the Pan Am aircraft was still trying to work out where they needed to get off the runway. The captain of the KLM aircraft applied take-off power believing the runway was clear, and reported over the radio, *'now at take-off'*, despite not being given a take-off clearance by air traffic control, and despite the Pan Am aircraft still being on the runway. The controller responded *'OK'*, non-standard phraseology, and likely due to misinterpreting the KLM radio message to mean they are ready for take-off. The Pan Am pilots were a little concerned about what they were hearing and radioed at this time, *'we're still taxiing down the runway'*, but this critical radio call was blocked, meaning neither the controller nor the KLM pilots heard the message. The message was blocked because of the way aircraft radios work. When two people try to speak at the same time on the same radio frequency, all that results is a shrill sound. The people making the radio call at the same time only hear their own voices and are unaware of the issue. The KLM aircraft raced down the runway, only seeing the other aircraft through the thick fog during the final few moments. The two aircraft tried to avoid each other but unfortunately collided.

To avoid the risk of confusion, the following words and phrases should be used during radio transmissions:

Acknowledge	Let me know that you have received and understood this message.

Affirm	Yes.
Approved	Permission for proposed action granted.
Break	I hereby indicate the separation between portions of the message.
Break break	I hereby indicate the separation between messages transmitted to different aircraft in a very busy environment.
Cancel	Annul the previously transmitted clearance.
Check	Examine a system or procedure.
Cleared	Authorized to proceed under the conditions specified.
Confirm	I request verification of… (clearance, instruction, action, information).
Contact	Establish (radio) communications with …. .
Correct	That is true or accurate.
Correction	An error has been made in this transmission (or message indicated). The correct version is…
Disregard	Ignore.
How do you read	What is the readability of my transmission?
I say again	I repeat for clarity or emphasis.
Maintain	Continue in accordance with the condition(s) specified or in its literal sense, e.g. maintain VFR.

Monitor	Listen out on (frequency).
Negative	No or permission not granted or that is not correct or not capable.
Out	This exchange of transmissions is ended and no response is expected.
Over	My transmission is ended and I expect a response from you. Note — OVER & OUT not normally used in VHF communication.
Read-back	Repeat all, or the specified part, of this message back to me exactly as received.
Recleared	A change has been made to your last clearance and this new clearance supersedes your previous clearance or part thereof.
Report	Pass me the following information ...
Request	I should like to know ... or I wish to obtain ..
Roger	I have received all of your last transmission. (Under no circumstances to be used in reply to a question requiring a READ-BACK or a direct answer in the affirmative or negative).
Say again	Repeat all, or the following part, of your last transmission.
Speak slower	Reduce your rate of speech.
Standby	Wait and I will call you.

Unable	I cannot comply with your request, instruction, or clearance.
Wilco	I understand your message and will comply with it.
Words twice:	a) As a request: Communication is difficult. Please send every word or group of words twice. b) As information: Since communication is difficult, every word or group of words in this message will be sent twice.

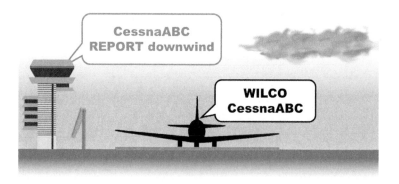

Figure 4.1: *Examples of standard words and phrases*

The list of standard words and phrases is considerable but highlights the importance of ensuring radio calls are clear and unambiguous. However, it is fairly easy to forget which word or phrase to use, especially when you are learning to fly. When this occurs just revert to plain English and keep the radio call brief. As you gain experience, these standard phrases and the other radio principles that are covered in this book will quickly become second nature.

You now have a good grasp of some of the key phrases that should be used when speaking on the radio. But how do you let air traffic control know who you are? Can you call yourself the little blue plane on the south ramp? Who are you talking with? In the next chapter, you will explore call signs, for you and the station you are talking to.

CHAPTER 5: **CALL SIGNS**

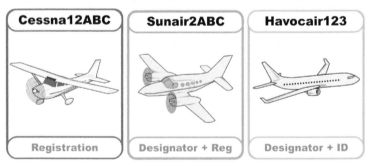

PILOT: Tower, please call me a fuel truck.
TOWER: Roger. You are a fuel truck.

Y ou are all ready to push down on the micro-phone button and make your first radio call, but there is one more important item that you need to consider – what do you call yourself? Your aircraft name is more commonly known as a call sign.

Cessna12ABC	Sunair2ABC	Havocair123
Registration	Designator + Reg	Designator + ID

Figure 5.1: *Examples of the three types of call signs.*

Call signs are very important as they ensure radio calls are delivered to the right aircraft. Improper use of call signs can result in pilots performing a clear-ance from air traffic control that was intended for another aircraft. But it's not just your call sign that needs to be considered, but also the call sign of the station you wish to talk to, such as Tower or Ground. You will see there are lots of different types of stations you may speak to on a flight.

Aircraft Call Signs

There are three different types of call signs. The most common and basic type of call sign is the aircraft registration marking, known as a type A call sign. Every aircraft has a unique registration – even large airliners. The aircraft registration will contain a series of letters and numbers, such as N47EB, G-ABC, or EC-ABC, just like a car registration plate. The first couple of characters indicate the country of registration. For example, N followed by digits indicates the aircraft is registered in the United States, G indicates the United Kingdom, and EC is used in Spain. When using your aircraft registration as a call sign, each character should be spoken correctly and individually, using the principles explored in chapter three.

Registration	Spoken
N47EB	November Fower Seven Echo Bravo
G-ABC	Golf Alpha Bravo Charlie

It is common practice to include the aircraft manufacturer or model in the prefix of a type A call sign (e.g. Cessna, Baron, Cherokee). This is to help air traffic control visualize the aircraft and its performance characteristics. When the aircraft manufacturer or model is included in the call sign, some countries allow the country prefix to be replaced, such as in the United States (e.g. N47EB becomes *Cessna47EB*), whereas, in other countries, the country prefix remains, such as the United Kingdom (e.g. *Cessna G-ABC*).

The second type of call sign is a mixture of a telephony designator and aircraft registration, known as a type B call sign. A telephony designator is normally used by commercial operators, for example, Sunair or Havocair. Type B call signs are more likely to be used by small commercial operators that do not have regularly scheduled services. This type of call sign includes the designator plus the last four characters of the aircraft registration. For example:

Designator	Registration	Call sign
Sunair	N47EB	Sunair47EB
Havocair	G-ABCD	HavocairABCD

The third type of call sign is the telephony designator plus the aircraft flight number (or flight ID), known as a type C call sign. This is primarily used by commercial operators which have flight numbers associated with each flight. Type C call signs may also include the word 'super' or 'heavy', which is determined by the aircraft wake turbulence category. For example:

Designator	Flight ID	Call sign
Havocair	937	Havocair937
Sunair	476	Sunair476 Heavy

It is very important your full call sign is transmitted when you make initial contact with a station. Some call signs, especially type A (aircraft registration), can be fairly lengthy, therefore there are strict procedures that allow some call signs to be abbreviated. To start with, a call sign can only be abbreviated once satisfactory communication has been established and the abbreviated call sign will not cause confusion. But how do you know when 'satisfactory' communication has been established? Satisfactory communication has been established once air traffic control has abbreviated your call sign (you are not allowed to abbreviate your call sign first).

Type A and B call signs can be abbreviated by taking the first character (or aircraft model/telephony designator) and at least the last three characters (last two in Europe). Type C call signs cannot be abbreviated. For example:

Full call sign	Abbreviation
Cessna N47EB	Cessna7EB or N7EB
Sunair G-ABCD	SunairBCD
Havocair476	No abbreviation

There should be no risk of confusion when a call sign has been abbreviated. This part can be a little tricky to judge as it largely depends on what other aircraft call signs are in use on the same frequency. The following case study highlights the dangers of call sign confusion;

On the 10th of May 2004, a Piper PA44 (Seminole) with a registration N304PA was conducting an IFR flight across California, maintaining 8,000 feet. The aircraft was one of five from the same company, flying the same route, each separated by 5 to 10 minutes. The aircraft directly ahead of N304PA was a Seminole with a registration of N434PA. The controller transmitted *'Seminole four papa alpha descend and maintain five thousand two hundred'*, which N304PA responded to and began descending. This clearance was intended for the first aircraft, but due to the same abbreviated call signs being used by the two aircraft, the controller was not aware the wrong aircraft had responded. The minimum safe altitude in the area was 7,700 feet, which N304PA descended below and collided with terrain.

The main reason for call sign confusion tends to lie with the final digits (e.g. ACF, JCF), parallel digits (e.g. 712, 7012), and anagrams (e.g. DEC, DCE).

Before leaving aircraft call signs, there are a few additions that can be made for some specific types of operations. For example, air ambulance flights will often have a prefix in their call sign to ensure they are given priority. In the United States, the prefix 'MEDEVAC' may be used (e.g. *MEDEVAC echo niner one*). The United States also has a call sign procedure to help student pilots, by including the term *'student pilot'* at the end of the normal call sign during the initial call. This helps the controller know which aircraft may need a little extra assistance (e.g. *november four three alpha bravo, student pilot).*

Station Call Signs

During a flight, you may need to speak to a range of different stations. A station is an air traffic service, such as a Tower or Ground. To understand the different stations, we first need to take a general look at the different types of airspace you may fly in. Airspace is roughly split into two areas; controlled or uncontrolled.

Controlled airspace is an area in which an air traffic service is provided. This is normally for sections of airspace with a reasonable amount of air traffic – especially commercial flights – and the air traffic service is primarily provided to ensure aircraft do not bump into each other. As a pilot, you need permission to enter this airspace and clearances are required (e.g. taxi or take-off clearance). Controlled airspace is normally found around busy aerodromes, in which a 'Tower' station will provide the air traffic service. A 'Ground' station may also be operating, which provides aircraft with ground movement instructions. There may also be a 'Delivery' station which is used by some IFR aircraft to obtain pre-departure instructions.

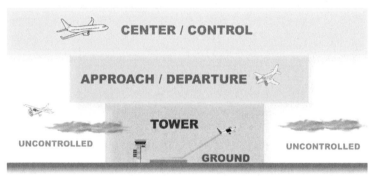

Figure 5.2: *A flight in controlled airspace may speak to a range of stations.*

Controlled airspace will also cover busy flight paths, which may include 'Departure' and 'Approach' – the intermediate section of airspace between take-off and cruise. It is fairly common that this intermediate stage is combined into a single Approach Control (e.g. controlling both approach and departure traffic). Air traffic services in the cruise phase of a flight are provided by control centers, which use the prefix 'Center' or 'Control'. These stations work seamlessly together.

Let's take a brief look at a flight in controlled airspace. You will start your flight speaking to a Ground station, which will provide clearances to taxi towards the runway. Prior to the runway, you will be transferred to Tower (at smaller airports, Tower may also provide ground movements). Tower will provide clearances that involve the runway (e.g. take-off clearance) and the initial airborne section, before transferring you to Departure (or Approach). Departure will then help deliver you to your cruising altitude and transfer you to a Center/Control station, which will provide clearances for the cruise section of the flight.

Each station will have two parts to their call sign; location or area name (e.g. Huston, Heathrow, Madrid) followed by the type of service (e.g. Control, Tower). For example, Huston Tower or Manchester Approach. Once satisfactory communication has been established and there is no likelihood of confusion, you can omit the name of the location or the call sign suffix (e.g. Huston Tower becomes Huston or Tower). The table on the following page shows the types of units or services that you may find during a flight. The table also includes some of the uncontrolled stations that will be explored next.

Unit or Service	Callsign Suffix
Area Control Center	Center or Control
Radar (in general)	Radar
Approach Control	Approach
Departure Control	Departure
Aerodrome Control	Tower
Surface Movements	Ground
Clearance Delivery (IFR)	Delivery
Flight Information Service	Information
Flight Service Station (FSS)	Radio
Aerodrome UNICOM	Unicom

Many flights remain outside of controlled airspace or only utilize controlled airspace for a portion of a flight (e.g. to transit). These flights will operate in uncontrolled airspace, which means no air traffic service is provided (e.g. no controller providing clearances or instructions). Radio calls are essential in uncontrolled airspace as they help keep everyone else aware of each other's position. Radio calls in uncontrolled airspace are primarily about exchanging traffic information, especially when approaching or departing an aerodrome.

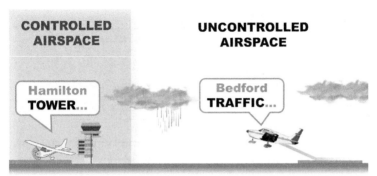

Figure 5.3: *Aircraft flying in uncontrolled airspace may use the prefix 'Traffic' or 'Radio' to report their position and intentions.*

The key to communicating at an aerodrome without an operating control tower is the selection of the correct common frequency, sometimes called a CTAF (Common Traffic Advisory Frequency). Some non-towered aerodromes provide some support for pilots, but this largely depends on the level of activity. Uncontrolled aerodromes can generally be split into three groups; Flight Service Station (FSS), UNICOM or MULTICOM. In all three environments, pilots need to regularly communicate their position and intentions and remain vigilant for other traffic (*traffic* is the term used to mean other aircraft).

Aerodromes with very low volumes of air traffic will have no ground-based advisory service available. That is, there is no extra pair of eyes or ears. Pilots at these aerodromes still need to make regular radio calls on the published frequency. This frequency is normally a common frequency used by similar aerodromes, sometimes called a MULTICOM. For example, in the United States, the common MULTICOM frequency is 122.9. The prefix at these aerodromes is normally *'Traffic'*, along with the aerodrome name.

> PILOT: Bedford Traffic, Cessna12ABC turning finals Runway 35, full stop, Bedford

Note the name of the aerodrome is repeated at the end of the transmission. This is important as other aerodromes in the general area may have aircraft also transmitting on the same common frequency. Transmitting the aerodrome name twice will minimize the risk of confusion.

When traffic volumes are higher at a non-towered aerodrome, a ground service in the form of a UNICOM or Flight Service Station (FSS) may be provided (some countries may use different names, such as an Aerodrome Flight Information Service [AFIS]). A UNICOM is partway between no extra help at all and an FSS. The UNICOM operator will provide advisory information, such as wind direction, altimeter settings, and the favored runway. When communicating directly with the UNICOM operator, the prefix *'Unicom'* is used (e.g. *Bedford Unicom, request field advisory*), but all other calls are made with the prefix *'Traffic'.* Some aerodromes may also offer an automated UNICOM, whereby selectable information is obtained by microphone clicks on the frequency. For example, 3 clicks of the microphone may provide the weather information.

The next level up for a non-towered aerodrome is a Flight Service Station (FSS). An FSS is an advisory service that can provide traffic information, briefings (e.g. weather, including at other aerodromes), file flight plans, and activate search and rescue. The suffix *'Radio'* is used when operating at an aerodrome with an FSS (or AFIS). For example:

> **PILOT:** Salisbury Radio, Cessna12ABC taking off Runway 13

Note even though a UNICOM or FSS may have a person providing information, they cannot provide clearances.

As we have just seen, there are many different stations that a pilot can communicate with. At times it may seem overwhelming, but the important thing to note is the person behind the radio is there to help.

CHAPTER 6: **BASIC FLIGHT RADIO TECHNIQUES**

TOWER: CessnaABC, we can't hear the beginning of your calls. PLEASE press the transmit button BEFORE you start talking!
PILOT: ...oger

When you learn to fly, speaking on the radio is often one of the most stressful aspects of the first flights. What if I say the wrong thing? What if air traffic control tells me off? You will see in this chapter that there are a few general techniques that can help you deliver a clear and concise radio call. Remember, clear communication is essential to reduce the risk of miscommunication between pilots and the controller. In this chapter you will also take your first proper look at parts of a radio call; should you say your call sign at the beginning or at the end of a transmission? You will also explore a very important type of radio call – the read-back.

It is finally time to talk on the radio. You are about

to push down on the microphone button for the first time, however, before talking, you should listen. That's right, listen to the radio frequency first. This is to ensure you are not stepping on someone else's radio call. Once you start using the radio regularly, you will see when a pilot and controller communicate, it is often like a game of tennis, with the radio call going back and forward between the pilot and controller. Interrupting this sequence of calls can cause confusion or frustration for all involved. Another issue is what happens when you speak at the same time as another person (pilot or controller). A major barrier in aviation communication is not the language but how the radio works. If you are talking to a friend on the phone, and you find they are rambling on, at any point you can butt in. That is, you can both talk on the phone at the same time. Radios in aircraft do not work like this but work more like trying to talk to someone on a walkie-talkie. Only one person can speak at a time, and there is no way for you to jump in until the other person has finished talking, even if you want to. This limitation can be frustrating for pilots and air traffic controllers and has contributed to misunderstandings, and even worse, as was seen in the Tenerife case study earlier in this book. When two people try to speak at the same time on the same radio frequency, all that results is a shrill sound (known as a heterodyne). The people making the radio call at the same time only hear their own voices and are unaware of the issue.

Another issue is a stuck microphone. There has been a number of cases in which pilots have accidentally left their microphone on. When a microphone switch/button is stuck, no one else can use the radio frequency (but can still hear every word you are saying!).

Once you have listened to the frequency, your next task is to think. Think about what you need to say. Remember, the airways can become very congested, especially in busy airspace, therefore it is important that radio calls are kept brief and to the point. When you do talk on the radio, speak at an even rate, not too fast nor not too slow. Often this comes with practice – in fact as you get more proficient, it becomes more common for you to rattle off the radio call too fast. Ensure you keep to the standard phraseology and words, much of which has already been covered in previous chapters. Avoid jargon, chatter, and slang.

Clearances

The position of your call sign can be confusing. Should you say your call sign at the start or at the end of a radio transmission? Let's start with the simple one. When air traffic control makes a radio call, the call sign of the aircraft they wish to talk to will be at the start of the transmission. This makes sense, you do not want to be waiting until the end of the radio call to work out if the message from the controller is intended for you or another aircraft. Even if the message is not intended for you, it is still worth listening to help build your situational awareness of the traffic in the area.

> **TOWER:** *CessnaABC*, taxi to holding point B1

This type of radio call is generally referred to as a clearance (i.e., authorized to proceed under the conditions specified). A clearance will often contain a clearance limit. This is the point to which you can proceed, but no further without another clearance. In the radio call above, the clearance limit would be *'holding point B1'*. Once you arrive at B1, you will need another clearance to proceed further (e.g. to line up on the runway or taxi to another holding point).

Another variation of a clearance is a 'conditional clearance'. This is a very efficient way of issuing clearances in busy airspace. As the name suggests, a condition has to be met before a clearance is valid. Consider a busy aerodrome, with lots of aircraft landing and taking off from a single runway. A controller can issue a conditional clearance to an aircraft waiting to line up before the aircraft on finals touches down. Basically, all the radio talk can be done early. For example:

> **TOWER:** CessnaABC, after the landing Twin Star has passed, line up and wait, behind

The clearance is to *'line up and wait'* but this is only valid once the condition has been met; *'the Twin Star has passed'*. You will also note sometimes a brief reiteration of the condition is added to the end (*behind*). When a conditional clearance involves an active runway (like the above example), it is important both the controller and pilot have correctly identified the necessary traffic. This can be met by the controller asking the pilot to identify the traffic before the conditional clearance has been issued (e.g. *report the Twin Star on final insight*). Once the condition has been met, there is no need for a lengthy radio call, you can simply line up and wait.

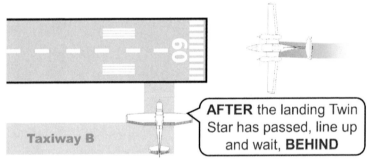

Figure 6.1: *Conditional clearances are an efficient way of issuing clearances in busy areas.*

Read-back

When you make a radio call, the position of your call sign in a transmission can roughly be split into two types; are you making a request (or providing information) or transmitting a read-back. When you are making a request or providing information (e.g. a position report), your call sign is placed at the start of a transmission:

PILOT: *CessnaABC,* ready at taxiway B1

When your transmission is a read-back, your call sign is placed at the end of the message. Read-backs are very important. A read-back is when you repeat all or a specific part of a message back to the controller, exactly as received:

PILOT: Cleared for take-off, *CessnaABC*

If the read-back is not exactly as the message received (or part of), air traffic control can pick up the miscommunication and ensure the pilot has the correct message. However, occasionally the controller does not recognize an incorrect read-back:

On August 8th 2009, a Piper PA32 (single-engine aircraft) was preparing to take-off from Teterboro Airport, New York. Just before noon, the pilot called ready for departure and the local controller cleared the aircraft for take-off, and instructed the pilot to make a left turn to the southeast (to avoid airspace) and maintain 1,100 feet or below. At the time, a helicopter was inbound to Teterboro Airport, which was advised of the departing traffic. After departure, the

PA32 pilot was instructed to contact Newark Tower on frequency 127.85. The pilot replied, *'one two seven point eight'* (missing the final 'five' in the frequency). The Teterboro controller failed to correct the read-back, as such the PA32 pilot switched over to an incorrect frequency. Shortly afterwards, it became apparent the PA32 and helicopter were on a collision course. The Newark controller issued an urgent instruction to the PA32 pilot to climb away from the conflicting helicopter traffic, however, the pilot was not on the correct frequency. The aircraft and helicopter failed to see each other and collided over the Hudson River.

Not all information requires a read-back, just the important parts. The following items are required to be repeated:

- ATC route clearances (IFR).
- Clearances and instructions to enter, land on, take-off from, hold short of, cross and backtrack on any runway.
- Runway in use, altimeter settings, SSR codes (transponder), level instructions, heading and speed instructions, and transition level.
- Taxi instructions, including holding short of the runway in use.
- Frequency changes.

TOWER: CessnaABC, wind 250 at 10 knots, cleared to land Runway 25
PILOT: Cleared to land Runway 25, CessnaABC

You will note in the above example, only the read-back items should be repeated by the pilot. Controllers will often provide additional information (e.g. wind and traffic information), but to avoid the airways becoming cluttered by unnecessary information, read-backs are only for vital information.

Frequency Changes

Throughout a flight, you will often need to change the radio frequency to speak to different stations, such as changing from the Ground to Tower frequency. It is important to only change frequency when instructed to by air traffic control. Frequency information is on the list of important items to read-back, and in an earlier chapter you saw the importance of saying the full frequency (including the term point or decimal). There are a few different ways in which a frequency change can occur. The most common way is when a controller uses the term CONTACT:

GROUND: CessnaABC, contact Hamilton Tower 118.5
PILOT: 118.5, CessnaABC

When the term CONTACT is used, this means you can switch to the new frequency straight away. You will note in the read-back, the call sign of the new frequency does not need to be repeated (Hamilton Tower). The important part is the frequency. Air traffic control may also use the term STANDBY:

> **GROUND**: CessnaABC, standby for Hamilton Tower 118.5
> PILOT: 118.5, CessnaABC

The term STANDBY means you can switch over to the new frequency, but wait for the new station to contact you. This is used when the new station is currently busy, and when they are free, they will contact you. Changing frequency at a specific point can also be utilized:

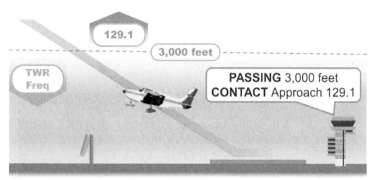

Figure 6.2: *Changing frequency at a specific point is an efficient use of the airways.*

TOWER: CessnaABC, when passing 3,000 feet contact Manchester Approach 129.1
PILOT: When passing 3,000 feet, 129.1, CessnaABC

In this type of frequency change, stay on your current frequency until the authorized change point has been reached (e.g. passing 3,000 feet). Before 3,000 feet, air traffic control can contact you on the Tower frequency, and after 3,000 feet you should be on the new frequency. MONITOR is another frequency change, whereby you are authorized to listen to another frequency, often for information that is being broadcasted (e.g. ATIS).

Test Procedure

Sometimes you may be concerned your radio is playing up. You cannot check the quality of your radio call by yourself, but you can ask for a RADIO CHECK. This is generally conducted with an air traffic service, but could also be conducted with another aircraft.

PILOT: Hamilton Tower, CessnaABC, radio check 122.9
TOWER: CessnaABC, reading you five

The number that the controller provides is in reference to the following readability scale:

1: Unreadable
2: Readable now and then
3: Readable but with difficulty
4: Readable
5: Perfectly readable

A station (e.g. Tower) may also ask you to check the readability of their radio, in which case you can reply using the same readability scale. The term HOW DO I READ is sometimes used to ask for a radio check as well.

Broadcast

The final type of general radio transmission is a broadcast. A broadcast is used by a station to send a message to all aircraft on that frequency. A range of information can be included in a broadcast, including changes in the weather or operating conditions (e.g. new ATIS or runway in use).

> **TOWER:** All Station, Hamilton Tower, Runway 26 now in use

You have listened on the frequency, thought about what you want to say, and now you know the general content of the message you wish to transmit. It is now time to push down on the microphone button and have a chat with the nearest air traffic controller. As you will see in the next few chapters, there are a range of radio calls to consider – from taxi to landing. But all follow the same communication phraseology that has been covered so far, with the same underlying goal of ensuring clear communication.

CHAPTER 7: **DEPARTURE**

TOWER: *Continue taxi holding position 26 South via Tango, check for workers along the taxiway*
PILOT: *Taxi 26 via Tango. Workers checked - all are working*

Y ou now have a call sign and have a good grasp of the range of common radio terminology. It is now time to talk on the radio. During a flight, you will make a range of radio calls and with practice, you will make these calls with very little effort, while conducting a range of other tasks in the aircraft. In this chapter, you will look at some of the common radio calls during the departure phase of a flight. In the next few chapters, you will look at different parts of the flight, from the cruise to the approach and landing. It must be stressed that when it comes to radio calls, there is no one size that fits all. Local radio procedures may be slightly different as radio calls can vary depending on aerodrome and airspace complexity.

Initial Call (Towered)

Let's start off with radio calls at a towered aerodrome (controlled). The first radio call will be the initial call. This is designed to establish contact with the station – like an initial greeting and will contain two parts; the call sign of the station being called, followed by your call sign. The initial call will sometimes include your position on the aerodrome surface (e.g. south ramp) and may include a request, if short (e.g. request taxi). This initial call is not just required for the start of the flight, but any time you contact a new station. Remember, on initial contact, your full call sign is required:

> PILOT: Hamilton Ground, Cessna12ABC
> **GROUND:** Cessna12ABC, Hamilton Ground

Sometimes there may be a short delay for the station to return your initial call as the controller may be performing other tasks. As a general rule, wait for at least 15 to 20 seconds before attempting your initial call again if you have not had a reply. Now that initial contact has been established, some call signs can be abbreviated. For example, you can now use the term 'Ground' or just 'Hamilton'.

Your next call will be longer and contain a number of parts depending on your intentions, but will generally contain the following items:

- Where you are (e.g. at the general aviation ramp)
- What you want (e.g. depart VFR to the south)
- The information you have received (e.g. with bravo).

PILOT: Cessna12ABC, south ramp, request taxi, VFR to Bedford, with bravo

There can be considerable variation to the above radio call. For example, if your aircraft type or model is not part of your call sign, you may include it in one of the first radio calls (e.g. *G-ABCD is a Cherokee on the general aviation ramp....).* It may also be normal practice to include the number of people on board (e.g. *3 POB*). POB is not required if a flight plan has been filed (as this information is in the flight plan).

The information part is referring to the ATIS (Automatic Terminal Information Service). ATIS is the continuous broadcast of recorded information that is available at some towered aerodromes, containing important meteorological information (e.g. wind speed, temperature) and operational information (e.g. runway in use). To obtain the ATIS, you will need to tune up a discrete radio frequency. This is a task you should conduct before making your initial call with a station.

For example, below is the type of information an ATIS may contain:

> *Hamilton information bravo. One three zero zero zulu. Expect ILS approach runway three six right. Surface conditions dry. Wind three five zero at eight. Visibility one zero. Ceiling four thousand five hundred broken. Temperature three four. Dew point two eight. Altimeter three zero one zero [or QNH 1020]. Advise on initial contact you have information bravo.*

ATIS content can vary and is unique to the aerodrome. As can be seen in the above example, the ATIS broadcast has a large amount of information. Having this information broadcasted on a separate frequency ensures the main frequencies do not become congested. The controller will want to ensure you have obtained the most recent ATIS information, which is why you report which ATIS you have based on the phonetic alphabet of the report. You would report '*with bravo*' for the ATIS information above. The ATIS is updated when there is a change in conditions. When this occurs, the phonetic alphabet identifier will also be updated (e.g. the next ATIS will be information charlie).

Taxi Clearance

The controller's reply to your initial request will likely contain taxi instructions. The taxi clearance provided can vary considerably depending on the complexity of the aerodrome. All taxi clearances contain a clearance limit. A taxi clearance is not a clearance to enter the runway or take-off. Sometimes a controller may use the term '*hold short*' to emphasize a clearance limit. If cleared to taxi *to* a runway, you are allowed to cross all other taxiways and runways on the way there, but you must hold short of the final runway.

> **GROUND:** CessnaABC, taxi to holding point G1 for Runway 27
> **PILOT:** Taxi holding point G1 for Runway 27, CessnaABC

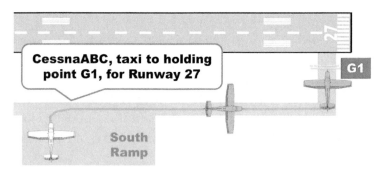

Figure 7.1: *An example of the taxi instructions provided by a Ground or Tower controller.*

Before entering the runway, you will be required to change to the Tower frequency (if you have not already).

GROUND: CessnaABC, contact Hamilton Tower on 123.4
PILOT: 123.4, CessnaABC

Take-Off Clearance

Once you are holding short of the active runway, the controller will provide the clearance to enter the runway. An important item to note in the following radio calls is the use of the word 'take-off'. The mis-understanding of the word 'take-off' can have significant consequences:

PILOT: CessnaABC, G1, ready for departure
TOWER: CessnaABC, Runway 27, line up and wait
PILOT: Runway 27, line up and wait, CessnaABC
TOWER: CessnaABC, Runway 27, cleared for take-off
PILOT: Runway 27, cleared for take-off, CessnaABC

There should be no confusion or ambiguity when an aircraft has been cleared for take-off (or not). As a result, generally, the only time air traffic control will use the phrase is when clearing an aircraft for take-off (e.g. *CessnaABC, cleared for take-off*) or to cancel the take-off clearance. Likewise, being such an important phrase, the aircraft is required to read it back (e.g. *cleared for take-off, CessnaABC*), which should be the only time an aircraft uses the phrase. *'Line up and wait'* is used by air traffic control to allow an aircraft to taxi onto the departure runway and wait. It is not permission to take-off, which will be issued separately.

At some aerodromes, the taxiway may not deliver you to the start of the runway (e.g. the taxiway is attached to the midway point of the runway, as shown in Figure 7.2). A back-track may be required to ensure you have sufficient runway for take-off. A back-track involves taxiing along the runway in the opposite direction to the take-off run, which must be communicated clearly:

> **TOWER:** CessnaABC, back-track and line up Runway 27

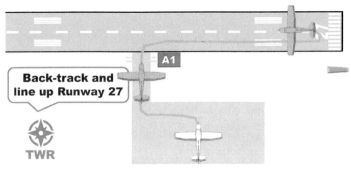

Figure 7.2: *A back-track may be required at some aerodromes, which involves taxiing along the runway in the opposite direction to take-off.*

Departure

Once airborne, you may require further frequency changes. This will be the case if moving from one type of controlled airspace to another (e.g. Tower to Approach). It is common both arrivals and departures are handled by a single approach control unit, as such, even though you are departing, you are communicating with Approach:

> **TOWER:** CessnaABC, contact Manchester Approach 118.75
> PILOT: 118.75, CessnaABC
> PILOT: Manchester Approach, Cessna12ABC, airborne, climbing 2,000 feet
> **APPROACH**: CessnaABC, Manchester Approach, report clear of the control zone
> PILOT: Wilco, CessnaABC

Occasionally a take-off may need to be aborted once the aircraft begins to roll. This may be initiated by the pilot (e.g. a warning in the cockpit) or the controller (e.g. they spot conflicting traffic taxiing onto the runway):

> **TOWER:** CessnaABC Stop Immediately, CessnaABC Stop Immediately
> PILOT: Stopping CessnaABC

You will note in the Tower radio call, the message was repeated twice to emphasize the unusual nature of the request (which is known as WORDS TWICE).

Non-Towered Radio Procedures

The radio procedures just discussed are for a towered aerodrome, which may or may not include a ground station. At a non-towered aerodrome (e.g. uncontrolled), the radio procedures are a little simpler. Typically, two radio calls are required during the initial departure phase; the first one just before taxiing and the second one before taxiing onto the runway:

PILOT: Bedford Traffic, Cessna12ABC, taxiing Runway 30, departing to the south

PILOT: Bedford Traffic, Cessna12ABC, lining up Runway 30

Uncontrolled aerodromes will not have an ATIS, but there may be other means to obtain weather and operational information. Some will have an automatic weather observing system (AWOS) or an automatic weather information broadcast (AWIB). These systems are similar to an ATIS, in that they broadcast information on a discrete frequency, but may not include operational information (e.g. runway in use). Non-towered aerodromes with a ground service (e.g.

UNICOM or FSS) will be able to provide pilots with information, such as runway in use or traffic information.

Surely once the busy take-off and departure phase has been completed you can relax? Sadly not. Some flights will continue into the cross-country phase, which may encounter complex airspace, requiring plenty of radio calls. Other flights may remain close to home, flying the traffic pattern (circuit), where regular radio calls are essential to keep everyone informed of aircraft movements.

CHAPTER 8: **CROSS-COUNTRY**

TOWER: You have traffic at 10 o'clock, 6 miles!
PILOT: Give us another hint! We have digital watches!

A considerable portion of a flight can be conducted in the cross-country or cruise phase. During this phase, you may go for long periods with very little contact with an air traffic service, as the controller may be able to 'see' you with the aid of your transponder or ADS-B. In this chapter, you will explore some of the different types of transponder modes and the corresponding radio calls. You will also look at flight plan related radio calls. A VFR flight plan can be a lifesaver and is strongly recommended for VFR pilots heading off on a cross-country flight.

Radar

Radar has been around for some time and is an extremely valuable tool to keep aircraft safely separated. To enter most airspace – especially busy airspace – an aircraft usually is required to be equipped with a transponder and increasingly an ADS-B. The information from transponders is also used for TCAS (Traffic Collision Avoidance System), which is fitted to some aircraft. This is why even in some uncontrolled airspace, a transponder may be required.

The first version of radar is known as 'primary' radar, which works similarly to weather radar. The ground-based radar unit sends out energy, and if an aircraft is flying nearby, some energy would bounce back to the radar site. On the radar screen, an aircraft would be spotted or be known as a 'target', but the controller would not know who this aircraft was. This type of radar system is still used around some busy airspace to identify non-transponder or ADS-B equipped aircraft.

The next advancement in radar is the secondary surveillance radar (SSR), which also goes by the name Air Traffic Control Radar Beacon System (ATCRBS). SSR is a big advancement in aircraft surveillance. This type of radar system requires the aircraft to be fitted with a transponder. In simple terms, the ground-based radar sends out a signal that is received by the aircraft transponder. The signal 'interrogates' the aircraft transponder, which sends back a coded signal to the radar unit, as shown in Figure 8.1. The ground-based radar site can work out the bearing and distance of the interrogated aircraft based on the returned signal. The information sent back to the radar site includes a four-digit number (squawk code) which helps the controller identify the aircraft. SSR has greater coverage compared to primary radar but is still limited by line-of-sight, therefore, can be of limited value in mountainous areas and areas a considerable distance from the ground-radar site.

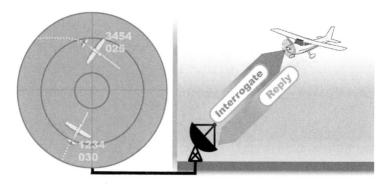

Figure 8.1: *The ground-based SSR 'interrogates' the aircraft transponder, allowing the aircraft position, squawk code (e.g. 1234), and in some modes altitude (e.g. 030 = 3,000 feet) to be displayed on a radar screen.*

There are three basic transponder modes, which represent the type of information the aircraft returns. Mode A is the most basic mode and simply returns a four-digit code, known as the 'squawk' code (e.g. 1234). When mode A is selected, the controller can see the aircraft's location on the radar screen and the four-digit code.

Mode C is the next level, which provides the same functions as mode A (location and squawk code), but also includes the aircraft's altitude. Altitude information is very useful as it helps the controller provide vertical separation between aircraft and is a key piece of information used in advanced TCAS (more on TCAS shortly). Note the altitude information is provided by the aircraft, not determined by the radar site. Mode S is the third mode, which includes all the information in Mode C, along with identification data. For example, the aircraft can transmit its registration and flight number.

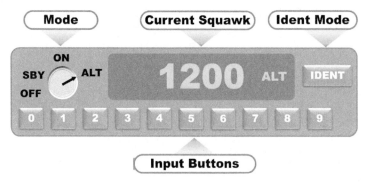

Figure 8.2: *An example of an aircraft transponder display.*

SSR is heavily used around the world, however, the next advancement in surveillance is slowly rolling out, which is known as ADS-B (Automatic Dependent

Surveillance-Broadcast). This is a shift from ground-based radar to satellite position reporting. Unlike SSR, the aircraft position is determined by GNSS (e.g. GPS) systems in the aircraft, which are relayed to the ground station for use by air traffic control. This has major advantages in areas that currently have limited SSR coverage, as expensive and large ground-radar units are not required, just a ground-based antenna that can receive ADS-B information from a passing aircraft, as shown in Figure 8.3. The standard mode is known as ADS-B Out, which means the aircraft can send out information (e.g. position), but cannot receive information. ADS-B In is generally an extra option found in more advanced units but enables aircraft to receive information, such as other aircraft position or weather information. This is a major tool that pilots can utilize to ensure they are operating with the most up-to-date information, allowing pilots to make informed decisions. Most controlled airspace still requires an operational transponder, as such, ADS-B is primarily a supplementary system at present, which can provide improved surveillance in areas with limited SSR coverage. Modern ADS-B units can operate both as a transponder and ADS-B system, and increasingly to enter airspace, not only is a transponder required but an ADS-B as well.

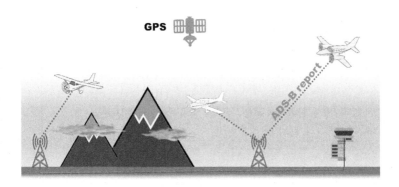

GPS

Figure 8.3: *ADS-B uses satellite position reporting (e.g. GPS) from aircraft.*

The different modes and functions of the transponder and/or ADS-B require specific radio procedures, most of which begin with the word SQUAWK. The following table shows a range of transponder/ADS-B related radio phraseology.

SQUAWK	Meaning
(number)	Select transponder/ADS-B code (e.g. SQUAWK 1234)
Ident	Select the IDENT feature on the transponder/ADS-B. The controller may be having trouble spotting you on their radar screen or you may have the same SQUAWK code as another aircraft. By pressing the IDENT feature, you flash up on the radar screen to aid in identification.

Standby	Switch transponder/ADS-B to standby function, which means the system is on but not transmitting.
Normal	Resume normal transponder/ADS-B operations, which may include normal mode selection and/or SQUAWK code.
Altitude/ Charlie	Select mode C (to allow altitude transmission)
Stop Altitude Squawk	Turn off automatic altitude reporting (e.g. select Mode A). This may be required if your transponder is transmitting an inaccurate altitude.

All transponder/ADS-B related radio calls require a read-back:

> **TOWER:** CessnaABC, Squawk 3420
> **PILOT:** Squawk 3420, CessnaABC

Aircraft may use a unique code assigned for the flight or a common code for the type of flight. For example, 1200 is the common VFR squawk code used in many countries (including the United States, Canada, Australia, and New Zealand). Sometimes you may be asked to SQUAWK VFR, which means select 1200. The transponder/ADS-B can also be used in an emergency.

There are three emergency codes:

Code	Meaning
7700	for emergency
7600	for radio failure
7500	unlawful interference

The transponder/ADS-B can also be a useful tool on the ground. Some aerodromes have systems that can track aircraft taxiing using transponder/ADS-B information, which is particularly useful in low visibility environments. Specific aerodrome procedures for transponder/ADS-B use will vary, but generally, aircraft should operate with the transponder in the altitude reporting mode any time the aircraft is positioned on any portion of the aerodrome movement area. This includes all defined taxiways and runways.

VFR Radar Services and Flight Following

With an operational transponder/ADS-B, a controller may be able to provide radar services. Radar services are regularly used by IFR flights, but it is important to note radar services can sometimes be provided to VFR pilots too. Some areas may have a service for VFR flights called *flight following*. Flight following means a controller watches your position and gives you reports of traffic in your area. You will be provided with a unique squawk code for this service:

PILOT: Manchester Approach, Cessna12ABC

APPROACH: Cessna12ABC, Manchester Approach, go ahead

PILOT: Cessna12ABC, over Bedford, request flight following

APPROACH: CessnaABC, Squawk 5449

PILOT: Squawk 5449, CessnaABC

Traffic information is often one of the most valuable types of information a pilot can receive during flight following. Traffic information will be provided in a standard format to help pilots visually locate other aircraft in their vicinity. Traffic information will first be described using the clock code relative to the aircraft's nose (e.g. 12 o'clock for straight ahead, 3 o'clock to the aircraft right). Next, distance in nautical miles, the direction of conflicting traffic, and type of aircraft if known. This system is not unique to flight following and will be used any time a controller passes on traffic information (e.g. in the traffic pattern):

APPROACH: CessnaABC, traffic 10 o'clock, 3 miles, crossing left to right

Controllers are normally more than happy to help but may be limited by the level of assistance they can provide if their workload is already high (e.g. controlling high volumes of traffic in congested airspace). When flight following is no longer required, you can inform the controller.

> **PILOT:** Manchester Approach, CessnaABC would like to cancel flight following
> **APPROACH:** CessnaABC, radar service terminated, Squawk VFR, frequency change approved
> **PILOT:** Squawk VFR, CessnaABC

For a VFR aircraft receiving radar assistance, it is important to note the controller is unlikely to be aware of the weather conditions that you are flying in. As such, you should keep the controller advised if their vectors (directions to fly) may result in you entering unfavorable weather conditions. If the clearance puts you in danger (e.g. a VFR aircraft entering IMC) you should inform the controller you are *'unable to comply'*. It may be possible for the controller to amend the clearance.

> **PILOT:** CessnaABC, unable to comply due cloud at 2,000 feet, request 1,500 feet

VFR Flight Plan

A flight plan is not required for most VFR flights but is strongly recommended, as it provides search and rescue (SAR) protection. If you have an emergency and need to make an off-field landing, the flight plan has critical information that can help rescuers find you swiftly. Ideally, a VFR flight plan should be filed before a flight, either online or over the phone. When you commence the flight, the flight plan needs to be 'opened', which is also known as activating the flight plan. If departing from an aerodrome with an air traffic service (Tower or FSS), the controller will 'open' your flight plan for you. However, if you depart from an aerodrome without an air traffic service (e.g. Tower off watch or a private strip), you will need to 'open' the flight plan with the nearest air traffic service (e.g. a nearby FSS):

> PILOT: Salisbury Radio, Cessna12ABC
> RADIO: Cessna12ABC, Salisbury Radio, go ahead
> PILOT: Salisbury Radio, open flight plan for Cessna12ABC, from Bedford to White Plains
> RADIO: Roger Cessna12ABC, flight plan activated

Many areas have a flight information service available for aircraft in the cross-country phase of a flight, especially VFR aircraft in uncontrolled airspace. As the name suggests, this service can provide information similar to a UNICOM or FSS. These information services can be used to obtain updated weather information or update a flight plan. Although not required, it is recommended you provide periodic position reports during a cross-country flight. This ensures your flight plan contains the most up-to-date information, which could be critical if a search and rescue operation is required:

PILOT: Cessna12ABC, 20 miles south of Hamilton, VFR flight plan, Bedford to White Plains

You also need to 'close' or cancel your flight plan. When landing at a towered aerodrome, your VFR flight plan will not automatically be cancelled, as the controller does not know if a particular VFR aircraft is on a flight plan. If you fail to report or cancel your VFR flight plan within 30 minutes of your filed estimated time of arrival (ETA), search and rescue procedures will commence.

Transiting

During a flight, you may stay within controlled airspace the whole time – which is often the case with commercial flights. In these cases, the controller will facilitate your transfer between different stations, just like you do when changing from a Ground to Tower frequency. You are required to follow the proper radio frequency change procedures that were discussed in an earlier chapter:

TOWER: CessnaABC, contact Seattle Approach 129.1
PILOT: 129.1, CessnaABC

Some flights will largely remain in uncontrolled airspace, but you may wish to transit controlled airspace for a portion of the flight. If you are planning to transit through controlled airspace, you will need to obtain a clearance. You must make your initial call in good time (normally at least 5 minutes before the airspace boundary):

> **PILOT:** Manchester Approach, Cessna12ABC
> **APPROACH:** Cessna12ABC, Manchester Approach
> **PILOT:** Cessna12ABC, VFR from Hamilton to Bedford, 15 miles south of Salisbury, enroute Bedford, request transit of airspace
> **APPROACH:** CessnaABC, cleared through airspace direct to Bedford, maintain 5,500 feet VFR
> **PILOT:** Cleared direct to Bedford 5,500 feet VFR, CessnaABC

Level Instructions

When operating in controlled airspace, you may need to change altitude/flight level occasionally. Like all other radio calls, it is important these messages are clear and with little risk of confusion. As noted in an earlier chapter, altitudes should include the terms 'hundreds' and 'thousands' when appropriate. Given the importance of altitude/flight level in maintaining safe separation with other aircraft, they require a read-back:

> **ATC:** CessnaABC, descend and maintain five thousand
> **PILOT:** Descend and maintain five thousand, CessnaABC

The clearance is clear and simple, head down to 5,000 feet now. Another level clearance that may be provided is 'descend at pilot's *discretion*, maintain five thousand'. 'Discretion' means the controller does not require the clearance (descent) to start when the clearance is issued, but the pilot can decide when they wish to descend.

Traffic Collision Avoidance System (TCAS)

Before we start our descent and explore the approach and landing radio calls, we will have a brief look at the Traffic Collision Avoidance System (TCAS). You may be wondering why we are looking at a cockpit system in a radio book. During a TCAS event is one of the few occasions in which sticking with an air traffic clearance may not be the best course of action:

On the night of July 1st 2002, a Tupolev 154 and a Boeing 757 were flying over Southern Germany. Both aircraft were flying at FL360 and on a collision course. Less than a minute before the collision, the controller realized the danger and instructed the Tupolev to descend to FL350 to avoid the collision. Seconds later, the TCAS in both aircraft activated, informing the Tupolev to climb and the Boeing to descend. Had both aircraft followed the TCAS instructions, the collision would have been avoided. The Boeing followed the TCAS instruction and descended, but could not inform the controller as the frequency was busy. The pilots of the Tupolev ignored their TCAS instruction to climb and descended as requested by the controller. A short time later the aircraft collided.

A mid-air collision is a major hazard for any flight. TCAS is a system that picks up the transponder signals of other aircraft and warns the pilot of a possible conflict. The most advanced versions of TCAS are normally only installed in larger commercial aircraft. More basic versions are also available, which may be called TCAS I or Traffic Awareness System (TAS), and are more likely to be found in a general aviation aircraft. A similar type of system is called Traffic Information Service (TIS). A TIS uploads data from a ground station of traffic in the area. These systems provide proximity warnings, but will not offer the pilot any recommended avoidance action.

The most advanced TCAS can provide pilots with not only information of conflicting traffic, but also how a pilot should respond. When two aircraft are on a collision course, the TCAS will first generate a Traffic Advisory (TA), warning of a possible risk of a collision. If the two aircraft get too close, a Resolution Advisory (RA) will be generated, which tells the pilot to either climb or descend to avoid the conflicting traffic, as shown in Figure 8.4. Both conflicting aircraft do not require a TCAS, in fact, the key component is both aircraft need an operating transponder. For example, on 29th September 2006, a Boeing 737 and a business jet collided over the Amazon rainforest at 37,000 feet, in clear weather. Both were almost brand-new aircraft with operating TCAS. However, a pilot in the business jet inadvertently turned their transponder off earlier in the flight. Without the transponder signal, neither aircraft TCAS system knew of the collision threat.

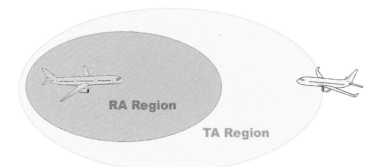

Figure 8.4: *A TCAS can provide a warning when an intruding aircraft is within the Traffic Advisory (TA) region and avoiding action within the Resolution Advisory (RA) region.*

When two TCAS operated aircraft are on a collision course, the TCAS systems ensure each aircraft takes different avoiding action (e.g. one aircraft is told to climb, the other to descend). For this reason, during a TCAS event, it is important to follow the RA. Pilots should respond immediately to an RA unless doing so would jeopardize the safety of the flight (e.g. descending into terrain). Due to the serious nature of the event, always fly the aircraft first, and then advise air traffic control as soon as practical:

> PILOT: Havocair, TCAS RA
> **APPROACH**: Havocair roger, report returning to clearance
> PILOT: Havocair, clear of conflict, returning to FL350

Deviations from a clearance should be kept to a minimum to comply with the RA, with most RAs requiring only a small deviation from an assigned altitude.

The cross-country phase of a flight can stretch for long periods. This phase may require multiple radio calls as you navigate through complex airspace. All flights need to arrive at an aerodrome at some point, so next we will look at radio calls in the final moments of a flight.

CHAPTER 9: **ARRIVAL**

PILOT: *Tower, Cessna123 is 5 miles northwest inbound for landing*
TOWER: *Cessna123, do you have information Hotel?*
PILOT: *Don't need information Hotel, I live in an apartment*

J ust like the radio calls conducted when departing an aerodrome, the arrival radio calls can vary depending on local procedures and the type of airspace you are entering (e.g. towered verse non-towered). However, what all radio calls will have in common is the traffic pattern (circuit) terminology, which is used by aircraft of all sizes. In this chapter you will explore the traffic pattern and see how most aircraft will utilize at least some part of the pattern when joining to land. You will also explore other radio calls that may be required in the final moments of a flight.

When approaching an aerodrome, the first few radio calls are similar to those at the start of the flight.

Initial contact should be made first, which is then followed by a call stating where you are, what you want, and what information (ATIS) you have received. Just like during departure, the ATIS will provide a rich source of information. Some pilots may report they *'have the numbers'* but this call is not referring to receipt of the ATIS, rather runway, wind, and altimeter information only. You should make your initial call in good time to allow for a planned entry into the traffic pattern or controlled airspace. Your position report is usually in reference to an easily identified point or landmark and should include the altitude of the aircraft:

> PILOT: Cessna12ABC, 10 miles northwest, 2,000 feet, inbound for landing, with charlie
>
> **TOWER:** Cessna12ABC, Hamilton Tower, cleared right base for Runway 29
>
> PILOT: Right base Runway 29, Cessna12ABC

Note pilots may also use the term *'inbound for landing'* or *'request joining instructions'*. Some aerodromes may have published arrival procedures, for example 'west arrival', in which case pilots should follow the set arrival procedure. Arrival procedures are often found at busier aerodromes – both towered and non-towered.

Traffic Pattern (Circuit)

As a student pilot, you will spend many hours going around the traffic pattern to help master your landings. The rectangular traffic pattern at the aerodrome is also called the circuit. The traffic pattern is not just for landing practice, but parts of the traffic pattern are regularly used when joining to land, as was seen in the joining instructions in the previous set of radio calls.

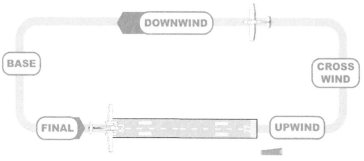

Figure 9.1: *A left-hand traffic pattern.*

Let's start by going around a traffic pattern, as shown in Figure 9.1:

- **Upwind:** The flight section that flies parallel to the landing runway in the direction of landing. This may be flown after take-off or during a go-around.
- **Crosswind:** At around 500 feet, a 90-degree turn to the left or right will be made onto the crosswind leg. The aircraft is usually still

climbing on this section.

- **Downwind:** The flight path that is parallel to the landing runway but in the opposite direction to the take-off direction. This is often flown at around 1,000 feet and is the longest of the legs. The term early or late downwind may be used in radio calls.
- **Base:** At the end of the downwind leg, a 90-degree turn is made towards the runway. The aircraft would normally be descending on this leg.
- **Finals:** When the base leg intersects the extended centerline, you turn to line up with the landing runway.

The altitudes mentioned may vary depending on local procedures and aircraft type. As a general rule, propeller-driven aircraft will fly downwind 1,000 feet above the ground, large aircraft will fly at least 1,500 feet (or 500 feet above the established pattern), and helicopters at 500 feet above the ground.

It is important to note which direction the aircraft turns in the traffic pattern. In Figure 9.1, the aircraft made all turns to the left, for example, a left turn from crosswind to downwind, and a left turn from base to finals. As a result, it can be said that this is a left-hand traffic pattern. You do not need to specify a left-hand pattern in radio calls (e.g. *downwind Runway 09*), although it is advisable to do so if you are at an aerodrome where the pattern direction is variable and there has been a recent change. Traffic patterns can also be right-hand (e.g. all turns to the right). This may be a common procedure at an aerodrome, to ensure the pattern does not fly over a populated area. As a result, 'right' should be included in radio calls, *'right downwind, Runway 23'*.

Joining the Traffic Pattern

When joining the traffic pattern, the controller will often require an aircraft to join directly onto one of the pattern legs (e.g. downwind, base) which will provide the most efficient entry:

> **TOWER:** CessnaABC join right base Runway 32, report on base
> PILOT: Join right base Runway 32, wilco, CessnaABC

Having joined the traffic pattern, you should make reports as required by local procedures. If you are planning on making a touch-and-go, low approach, or go-around, it is useful to advise the controller as part of the downwind call (*CessnaABC, downwind, touch-and-go*). Some aircraft may be *'cleared for the options'*. This clearance is used for dual training flights, which gives the pilot the option to make a touch-and-go, low-approach, missed approach, stop-and-go, or full-stop landing. This type of clearance can be very beneficial in training situations. If you do not state your intentions, the controller will assume you intend to land. The runway designator is not normally included with a position call (e.g. *downwind Runway 29 to land*), but if there is any possibility of confusion (e.g. more than one runway in use), you should include the designator:

> PILOT: CessnaABC, downwind Runway 09

If the traffic is light, you may be provided with a landing (or touch-and-go) clearance in the downwind position, in which case no further radio calls would be required. However, if other traffic is due to land or take-off before you, a sequence number may be provided:

> **TOWER**: CessnaABC, number two, follow Cherokee on base
> PILOT: Number two, traffic in sight, CessnaABC

When a sequence number is provided, you are expected to follow the aircraft in front and make allowance for separation. To help accommodate sequence, the controller may require you to adjust the traffic pattern, as shown in Figure 9.2. For example, if another aircraft needs to depart first, you may be asked to make an orbit (360-degree turn, taking 2 minutes), extend downwind, or join a long final (greater than the standard 4nm finals):

TOWER: CessnaABC, orbit left due traffic on the runway

PILOT: Orbiting left, CessnaABC

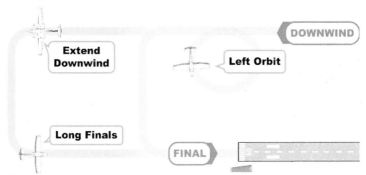

Figure 9.2: *To help accommodate sequencing, the controller may require an aircraft to adjust the traffic pattern.*

If a landing clearance has not already been provided, the clearance would normally be provided on finals. At some aerodromes, it is standard practice for the controller to include *'check wheels down'* when you are cleared to land. This is a standard thing to say whether your wheels are down or not. At aerodromes with parallel runways (e.g. Runway 27 left or Runway 27 right), the word 'left' and 'right' should be included with the runway designator (e.g. *cleared to land Runway 27 left*):

TOWER: CessnaABC, Runway 32, cleared to land

PILOT: Runway 32, cleared to land, CessnaABC

At any point in the landing phase, a go-around can be performed. A go-around may also be called an 'overshoot'. This may be initiated by the pilot (e.g. flying too fast or high) or by the controller. If time is available, the controller may add the reason for the go-around instruction:

> **TOWER:** CessnaABC, go-around, aircraft on the runway
> **PILOT:** Going around, CessnaABC

When conducting a go-around, unless otherwise instructed, climb back up to the traffic pattern altitude and continue around the pattern. Particular care should be taken to ensure the crosswind turn is conducted at the normal position, and not too early, which could result in turning in front of other traffic:

> **PILOT:** CessnaABC, going around
> **TOWER:** CessnaABC, roger, report downwind
> **PILOT:** Wilco, CessnaABC

After landing, taxi instructions will be provided, often towards the end of the landing roll. You may be asked to vacate to a holding point and transferred to the Ground frequency for further taxi instructions. To avoid confusion with clearances to land or take-off, you should use the term '*vacated*' when taxiing off the runway. The runway is vacated when the whole aircraft is beyond the relevant runway holding position. Make sure you understand the holding point and taxiway markings to ensure you do not stop before the runway holding point:

> **TOWER**: CessnaABC, taxi to the end, report runway vacated
> PILOT: Taxi to the end, wilco, CessnaABC
>
> PILOT: CessnaABC, runway vacated

Non-Towered Radio Procedures

The radio procedures just discussed are for a towered aerodrome. At a non-towered (uncontrolled) aerodrome the radio procedures may be a little simpler but extra care is required when joining to determine which runway is in use. You are probably not alone out there. The purpose of calls at non-towered aerodromes is to make others aware of your position, and for you to be aware of others. There could very well be people in the pattern who don't even have radios.

You should listen out on the necessary frequency well before joining to help build a mental picture of the location of traffic in the area. Non-towered aerodromes with a ground service (e.g. UNICOM or FSS) will also be able to provide pilots with information, such as runway in use or traffic information. The first radio call should be conducted at least 10 nautical miles from the aerodrome. This is a general call indicating your location and intentions:

PILOT: Bedford Traffic, Cessna12ABC, 10 miles to the north, inbound at 2,000 feet

The next call is typically when joining one of the traffic pattern legs, such as, downwind, base or straight in (long finals):

PILOT: Bedford Traffic, Cessna12ABC, joining left downwind for Runway 32

Once established in the traffic pattern, the calls you make will be dictated by traffic levels and local procedures. Some experienced pilots prefer that pilots do not jam up the airways with every single detail of their pattern. Use your judgment, and keep the calls short. But generally, at least a call on downwind, finals, and exiting the runway:

PILOT: Bedford Traffic, Cessna12ABC, turning finals for Runway 32

PILOT: Bedford Traffic, Cessna12ABC is clear of the active

You have now completed a full flight, from taxi, take-off, cross-country, and joining. Each flight conducted will have a unique set of radio calls, but all have the same common theme. The radio is an essential tool to help pilots safely conduct a flight. With practice, radio calls will become second nature and you will become well adapted to variations in local radio

procedures. From time-to-time, you may hear unfamiliar radio calls, these may be from an IFR flight –
which can be conducted by aircraft of a range of sizes.
You will see in the next chapter that although most
flights share similar radio calls, there are a few differences which all pilots should be aware of.

CHAPTER 10: **IFR RADIO CALLS**

Pilot: Do you know it costs us two thousand dollars to make a 360 in this airplane?
Controller: Roger, give me four thousand dollars worth.

During a flight, you may hear some unfamiliar radio calls, such as an aircraft cleared for a SID, ILS, or VOR. These radio calls belong in the world of IFR flying. IFR radio calls may be new and confusing at first. For those unfamiliar, IFR stands for instrument flight rules, which means the aircraft and the pilot can fly in instrument meteorological conditions, such as poor visibility and clouds. Most of the radio calls discussed so far are used in both VFR and IFR operations, however, IFR flights have some unique radio calls. This chapter is not aimed at making you an expert in IFR operations, but to provide a general overview of some IFR radio calls, with the main focus on providing VFR pilots with a little understanding of some of the strange radio calls they may hear.

IFR Departure

The first IFR radio call generally occurs before the flight even begins, known as a 'route clearance'. Unlike VFR operations, a flight plan must be submitted for almost all IFR flights. This flight plan is more detailed than a VFR flight plan. The route clearance (also called pre-departure clearance) does not actually clear an aircraft to start moving, but rather sets up key aspects of the flight; such as a general confirmation of the flight plan (along with any changes), providing initial routing requirements (e.g. departure instructions) and sometimes assign a unique transponder code for the flight. A route clearance can be quite lengthy and as a result, the transmission may be done on a 'Clearance Delivery' frequency. The route clearance may also be obtained later on, during taxi in some cases. It is important to note a route clearance is not an instruction to take-off or enter an active runway. A route clearance can vary widely, but an example is found below:

> **GROUND:** Cessna12ABC, cleared to Hamilton via Victor Eight, Intercontinental one departure, maintain eight thousand

The next unusual IFR radio call is an engine start-up request. Unless a local procedure requires it, most VFR flights can start their engine(s) as they wish. But IFR flights, primary at towered aerodromes, often need to request a start:

> **PILOT:** Hamilton Ground, Havocair123, stand 24, request start-up, with bravo
> **GROUND:** Havocair123, start-up approved
>   Or
> **GROUND:** Havocair123, expect start-up at 35 [time 35]

The start-up request is not for safety reasons – just like any flight, an IFR pilot must visually check the area is clear before starting the engines. But rather, the start call is for ATC planning and to avoid excessive fuel burn by aircraft delayed on the ground. If significant delays are expected, an IFR flight may be asked to wait before starting.

The taxi radio calls are much the same as a VFR flight, with the next IFR call occurring during departure. An IFR flight can depart an aerodrome following a number of different procedures. A VFR departure is generally fairly simple, such as, *'depart to the west, 2,000 feet and below'*. During a VFR flight, the pilot must remain in visual conditions and navigate visually (e.g. look outside to make sure they do not bang into anything). An IFR flight needs to be able to safely operate even if the pilot cannot see outside. As a result, IFR flight procedures are more structured.

If it is a nice sunny day, an IFR flight can depart visually like a VFR flight, but if the weather conditions are poor, an IFR flight may need to follow an IFR departure, such as a standard instrument departure (SID) or radar departure. A SID is a set procedure that is published in the aerodrome charts (e.g. BUDON TWO departure; climb on radial 090 to 2,000 feet, then turn left onto radial 120 and climb to 5,000 feet....). A radar departure may be used in airspace with sufficient radar coverage and is an efficient way for an aircraft to depart. A radar departure will involve the controller issuing radar vectors (heading instructions), which may also include speed requirements.

Figure 10.1: *An IFR flight needs to depart safely, even in poor visibility.*

The cruise phase of an IFR flight generally consists of operating in controlled airspace, such as an area control (center). These areas can vary in size and can range from a simple one-person unit to a large sophisticated center. IFR aircraft will often be transferred from one controlled area to the next. But do note, IFR flights are not limited to controlled airspace, however, larger aircraft flying IFR flights will generally remain in controlled airspace for the majority of the flight.

IFR Approach

As an IFR flight approaches its destination, it may be required to conduct a number of IFR procedures that may be unfamiliar to VFR pilots. The first is flying a few laps of a hold. A hold is like a queuing procedure for IFR flights, and may be referred to as *'flying a race track pattern'*. A hold is a safe place for an IFR aircraft if they need to wait their turn to fly an approach or wait for the weather to improve. The hold is generally fairly high, well above the normal traffic pattern (to remain safe above terrain), and can be flown at set places near an aerodrome, including directly overhead.

The next IFR procedure is the approach, which will vary depending on the weather conditions and approach aids available at the aerodrome. If the weather conditions are suitable, an IFR flight may join using similar procedures to a VFR flight, such as joining a leg of the traffic pattern:

> PILOT: CessnaABC, over Hamilton VOR 3,000 feet, field insight, request visual approach
> **APPROACH:** Cleared visual approach Runway 24, join right downwind, number 1, contact Tower 118.7
> PILOT: Cleared visual approach Runway 24, right downwind, 118.7, CessnaABC

If the weather is poor, an IFR flight will use approach aids to help guide the aircraft down to the runway safely. These aids can vary – the most precise is an ILS approach (Instrument Landing System), but others may include VOR, NDB or GPS approaches. Again, a range of radio calls are made during these IFR approaches, but what they all have in common is that they set the IFR aircraft up for a long final (often 8 to 12 nm).

> **APPROACH**: Havocair123, descend to 4,000 feet, expect ILS approach Runway 16

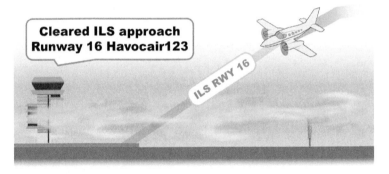

Figure 10.2: *An ILS approach can guide an appropriately equipped aircraft to the runway in poor visibility*

IFR to VFR

During a flight, a pilot may wish to change from an IFR to a VFR flight. This procedure may suit some pilots flying smaller aircraft. For example, if the weather is below VFR weather conditions at the departure aerodrome (e.g. fog), the pilot can depart IFR, and once the weather conditions improve, switch to VFR. This procedure is not normally used by large commercial aircraft that remain IFR for the whole flight, even if the weather is perfect.

> **PILOT:** Hamilton, CessnaABC, request cancel my IFR flight, proceeding VFR

In the last few chapters, you have covered a range of normal radio calls. But what happens if a flight is not normal? What if an emergency arises or your magic radio box gives up on you? Although such situations are rare, it is important to be prepared, which you will explore next.

CHAPTER 11: **RADIO FAILURE**

TOWER: *Cannot read you, say again!*
PILOT: *Again!*

I magine you have just entered a busy stretch of controlled airspace. There is a constant stream of radio chatter in your ear, but suddenly the radio is completely silent. You cannot hear a single radio call, despite dozens of aircraft flying in your vicinity. You look up at the radio and to your horror, where bright digits of radio frequencies were moments ago, it is now completely blank. First and most importantly, do not panic. The aircraft is not about to drop out of the sky. As you will see in this chapter, there is a range of procedures in place for such a situation to help keep everyone safe. It is impossible to cover all possible scenarios in which a radio failure could occur. But with good pilot judgment and a general understanding of the radio failure procedures, you'll be able to choose the appropriate action.

You may have heard the classic phrase; aviate, navigate, communicate. This simple list is a pilot's order of priority during any flight – including during an emergency, which you will visit in the next chapter. Aviate first – that is, make sure you are flying the aircraft. A radio failure can be stressful, especially if it occurs during a busy flight phase. If your aircraft is equipped with an autopilot, a radio failure is a perfect situation to utilize it, as the autopilot can reduce your workload, allowing you to concentrate on other tasks – like trying to fix the radio. Next, navigate. Make sure you remain aware of the direction that the aircraft is going. All too often, a minor distraction has resulted in pilots failing to keep an eye on where they are heading, sometimes leading to controlled flight into terrain. You should also be extra vigilant for other traffic in the area. You should be regularly looking out, but when the radio fails, this task becomes even more important. And finally, communicate. This is a little more challenging when the radio fails, but there are some communication methods that can still be deployed, which you will see shortly.

Your next task is to complete a quick investigation to determine if you can solve the radio failure. Things you can do and check include:

- Is your headphone jack plugged in correctly? Headphone jacks (plugs) can easily come loose if you have unintentionally knocked them out of place.
- Have you selected the correct frequency and

set the volume level correctly? It is easy to accidentally knock one of the frequency knobs. Changing the frequency a fraction can lead to a problem. Be especially wary if you have just changed frequency. If you are having trouble contacting the new station, you may have the incorrect frequency.

- Is your headphone working correctly? If you have a spare headphone or are flying with someone else, check to see if another headphone is working.
- Is the avionic/electrical switch turned on? In some aircraft, there is a separate switch for the radios. Switches can be turned off accidentally.
- Have the circuit breaker/fuses popped out?
- Are you flying in an area in which radio communication may be difficult, such as in a radio shadow of a mountain? Flying higher – if it's an option – may help.

If all this fails to identify the issue, there is one more thing to check. How do you know if your radio is working fine but the controller's radio has failed? Although rare, it does happen from time-to-time. You can try calling another aircraft on the frequency or tune in a nearby ATIS. A third possibility, is some parts of your radio are working, but not all. For example, you can receive (hear the controller), but not speak. Or you can talk but not receive. You will see once you go through some of the following radio failure procedures, these types of situations can be managed reasonably well.

You have now established you have a full radio failure, the first action you can do is make air traffic control aware of your situation with your transponder/ADS-B. Most aircraft are fitted with one, especially when operating in busy airspace, and squawking 7600 will alert the controller of your situation. Still make your normal radio calls, which is known as transmitting blind. This is just in case the controller can still hear you (receiver failure), and prefix radio calls with '*transmitting blind*':

> **PILOT:** Transmitting blind, CessnaABC, downwind Runway 23 to land

What you do next will depend on where you are (e.g. in the traffic pattern, controlled airspace or uncontrolled) and whether you are in visual meteorological conditions (VMC) or instrument meteorological conditions (IMC).

VMC Radio Failure

If the radio failure occurs when operating in visual meteorological conditions, make sure you remain in visual conditions. If you are an IFR flight and the failure occurs when operating in VMC conditions, if possible, remain in visual conditions. You should then proceed to the nearest 'practicable' aerodrome. What 'practicable' means depends on your aircraft type and current location.

If the failure occurs in controlled airspace, obey the most recent clearance and avoid high traffic areas. If practicable, divert to a non-towered (uncontrolled) aerodrome. If it is not practicable to divert to a non-towered aerodrome, only continue in controlled airspace if you are confident you are aware of the location of nearby traffic. Some situations are a little easier to judge, for example, if the radio failure occurred during the downwind leg of the traffic pattern, the most practicable aerodrome to land at would be the one you are currently flying at.

When approaching to land at a towered aerodrome, keep an eye out for light signals which provide a source of communication, as shown in the following table:

Signal	In Flight	On Aerodrome
Steady Green	Cleared to land	Cleared to take off
Steady Red	Give way to other aircraft and continue to circle	Stop
Green Flashes	Return to land	Cleared to taxi
Red flashes	Aerodrome unsafe – do not land	Taxi clear of the landing area in use
White flashes	Land at this aerodrome and proceed to the apron	Return to starting point on this aerodrome
Red and green flashes	Danger – be on the alert	
Red pyrotechnic	Notwithstanding any previous instruction, do not land for the time being	

You can acknowledge the light signal by either rocking your wings in flight (during the day) or flashing the aircraft lights on and off (at night).

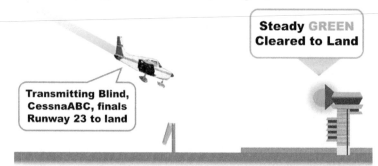

Figure 11.1: *Light signals from a tower can help communicate with an aircraft with a radio failure*

If the failure occurs in uncontrolled airspace, remain in uncontrolled airspace and land at the nearest suitable aerodrome. An overhead re-join may be required when joining a non-towered aerodrome to help determine runway direction and location of traffic. If an aerodrome has a published arrival procedure, follow the procedure. Not all aerodromes have the same radio failure procedures. If your destination aerodrome publishes radio failure procedures, you should make yourself familiar with these when planning the flight.

Much of these procedures are based on not being able to contact air traffic control at all. But in recent years, most people carry a handy device that can help overcome these communication issues – the mobile phone. Provided you have reception, you can simply call up the air traffic unit. The number is often published with the aerodrome information. It can be difficult to talk on the phone in a noisy aircraft, but at least this method provides some means of communicating.

IMC Radio Failure

In some ways, a communication failure during a flight in instrument metrological conditions (IMC) is a little more straightforward. If you are flying in IMC, you will be conducting an IFR flight. Most IFR flights must submit a detailed IFR flight plan. As a general guide, if an IFR flight encounters a radio failure, the pilot should follow the details in the submitted flight plan. The pilot knows what this plan is, and importantly, so does air traffic control.

The route an IFR flight should fly during a radio failure may be different from the one nominated in the flight plan. For example, you should continue on any route assigned in the last clearance first, then the route filed in your flight plan. The altitude flown should be at least the minimum en route altitude (to make sure you remain safely above terrain), followed by the last assigned altitude by air traffic control.

Speechless Technique

Another possible radio failure combination is when the radio transmitter is working but the microphone is not. This means when you hold down the microphone button, speech is not transmitted, but an unmodulated transmission may be heard – which sounds like a squelching noise. This can be used in a morse code kind of way, provided the controller provides YES or NO style questions. Can you see the aerodrome? Have you reached the nominated position?

To reply to these questions, you can use the following technique:

Meaning	Activate transmitter
Yes or Roger	once
No	twice
Say Again	three times
At nominated position	four times

A radio failure can be a very stressful situation. But just remember, the aircraft is still flying fine and there is a range of procedures to help keep you and everyone else safe. Once safely on the ground, phone the nearest air traffic unit to inform them what happened. Ensure your radio equipment is properly maintained, and if a problem is suspected with the radio, have it checked out before flying.

CHAPTER 12: **EMERGENCY**

PILOT: *Tower, CessnaABC, student pilot, I am out of fuel.*
TOWER: *Roger CessnaABC, reduce airspeed to best glide!! Do you have the airfield in sight?!?!!*
PILOT: *Uh...tower, I am on the south ramp; I just want to know where the fuel truck is.*

When an emergency arises during a flight, pilots should use all the resources they have available. Some resources come to mind easier than others, such as help from a second pilot or the autopilot. But one of the most useful resources that may be available, especially for a single-pilot, is air traffic control. Pilots should seek assistance from air traffic control whenever they are in doubt about the safety of the flight. An early call often prevents the situation from becoming more serious.

Utilizing air traffic control during an emergency is an effective way to bring an emergency to a safe and successful outcome. In fact, not using air traffic control can be disastrous, such as not letting them know you are low on fuel, and therefore needing priority:

On January the 25th 1990, a Boeing 707 was flying from Bogota to New York. The flight departed with plenty of fuel on board, however, after being placed in three holding patterns during the flight due to congested airspace, the aircraft became critically low on fuel. In the third hold, the pilot reported '*ah well I think we need priority*'. The controller asked how long they could hold for, with the pilot replying they could hold for five or more minutes. A short time later, they began their approach into New York, JFK airport. The first approach into JFK was aborted due to poor weather (windshear). During the go-around, the captain asked the first officer to '*tell them we are in emergency*'. The first officer told the controller that '*we'll try once again, we're running out of fuel*'. The dire situation was not recognized by the controller as the pilot did not use the word '*emergency*', either in this radio call or previous calls. The controller began sequencing the aircraft for another approach, however, as they were unaware the aircraft was running out of fuel, directed them further away from the airport to follow another aircraft. As the aircraft was preparing for the second approach, it ran out of fuel and crashed 20 miles (26km) from the airport.

As stated in the previous chapter, pilots need to aviate – navigate – communicate, and in that order. Once the first two items are under control, it is time to touch base with the nearest air traffic service. Emergencies are split into two categories; an urgency or a distress condition.

Distress

The highest level of emergency is a *distress condition*, and as such, will be given the highest priority by the controller. A distress is a condition of being threatened by serious and/or imminent danger and that requires immediate assistance. Just a few examples include an engine failure or a fire (engine or otherwise). During an emergency, air traffic control can be hugely helpful; they can ensure you are given priority to land, provide directions to the nearest aerodrome and ensure emergency services are available on landing. If necessary, they can relay messages, for example, to a ground engineer to help solve a technical issue.

The word 'MAYDAY' is spoken at the start of the radio call to identify a distress message. To help the controller give you maximum assistance, the emergency transmission should contain as many of the following items as possible, ideally in the order shown. However, you may need to change the radio call for your specific situation and the time available:

- MAYDAY, MAYDAY, MAYDAY
- Name of the station addressed
- Call sign of the aircraft
- Nature of the emergency
- Intentions
- Position (including level and heading)
- Any other useful information

PILOT: MAYDAY MAYDAY MAYDAY, Hamilton Tower, CessnaABC, engine on fire, making a forced landing, 20 miles south of aerodrome, passing 3,000 feet, heading 360

TOWER: CessnaABC, Hamilton Tower, roger mayday, wind at Hamilton 350 degrees 10 knots

A distress call should normally be transmitted on the frequency in use at the time. For example, if you are currently in contact with Tower, transmit the distress call on the Tower frequency. However, in some cases, it may be appropriate to transmit the distress call on another frequency. For example, if flying in uncontrolled airspace, it may be appropriate to tune up an air traffic service (e.g. a nearby Approach Control). If you are having trouble contacting anyone that can provide assistance, the emergency transmission can be made on 121.5 MHz – the international aeronautical emergency frequency. An emergency call transmitted on 121.5 MHz may be heard by a passing airliner (who may have 121.5 selected on one of their radio units), that can relay your distress message if required. The final important task you can do is squawk 7700, which if you are flying in an area with radar coverage, can alert air traffic control that you have an emergency.

Urgency

The next level of emergency is an *urgency condition*. An urgency is a condition concerning the safety of an aircraft or other vehicle, or of some person on board or within sight, but which does not require immediate assistance. This is a fairly wide-ranging definition and covers a range of situations, but basically, the main thing you are likely to want is priority. For example, if your fuel is low, you can transmit an urgency message – which means you are not in a distress situation at the moment, but if you are delayed trying to land, you might be. Other situations that would come under 'urgency' include a person is sick on board the aircraft and as a result, you need to land as soon as possible and medical help may be required on landing. Note the urgency call also covers 'other vehicles', that is, if you see an emergency with someone else, you can make an urgency call.

The radio transmission is very similar to the distress radio call, but starts with the words 'PAN PAN'. The urgency radio call should include as many items as possible as highlighted in the distress call. The urgency radio call should also be transmitted on the frequency currently in use and squawk 7700.

> PILOT: PAN PAN, PAN PAN, PAN PAN, Hamilton Tower, CessnaABC 10 miles north at 2,000 feet. Passenger with suspected heart attack, request priority landing

> TOWER: CessnaABC, Hamilton Tower, number 1, cleared straight-in approach Runway 18, wind 190 degrees 10 knots, ambulance alerted

When another aircraft has an emergency, especially in distress, it is very important that the aircraft with an emergency is able to communicate freely. As a result, radio silence may be imposed on all or some traffic on the frequency. Aircraft on the frequency should maintain radio silence until advised:

> TOWER: All stations, Hamilton Tower, stop transmitting, mayday

> TOWER: All stations, Hamilton Tower, distress traffic ended

When you hear a distress or urgency call, the first thing you should do is remain silent to allow air traffic control to acknowledge the aircraft with the emergency. But what if the emergency traffic is unable to contact an air traffic service? The aircraft in trouble may be low in a valley, which is blocking their emergency transmission to a nearby Tower. In this situation, you can relay the distress or urgency radio call, as touched on earlier with the use of the emergency frequency121.5 MHz.

Navigational Assistance

A pilot may come across a situation that does not quite reach the urgency or distress threshold but may need a little assistance. Being unsure of your position might fall into this category, and air traffic control may be able to provide assistance. If in doubt, a pilot should not be hesitant to make a PAN PAN call, especially if you are also experiencing another problem (e.g. low on fuel). A radar-equipped air traffic control unit can provide radar and navigational assistance to VFR aircraft provided the aircraft is within radar coverage (and the aircraft has a transponder/ADS-B unit):

> PILOT: CessnaABC, 2,000 feet, unsure of my position, request heading to Bedford
> APPROACH: CessnaABC, fly heading 160, your position 6 miles north of Bedford
> PILOT: Heading 160, CessnaABC

An air traffic facility may still be able to provide assistance to an aircraft outside of radar coverage using very high frequency direction finding (VDF). An aircraft does not need any special equipment, as the VDF system uses the aircraft radio transmission to work out which direction the aircraft is. The controller can then provide a range of navigational information, such as magnetic heading to steer (no wind) to reach the VDF station. When requesting a bearing, you should end your transmission by repeating your call sign. This lengthens your transmission, thereby helping the VDF station work out the aircraft bearing:

> PILOT: Cessna12ABC, request VDF steering to Salisbury, Cessna12ABC
> TOWER: Cessna12ABC, heading to Salisbury 060
> PILOT: 060, Cessna12ABC

A few important items to note. VDF is only available at some locations. The accuracy of the bearing can vary. For example, 'class bravo' may be added to a radio call, which means the heading is accurate within +/- 5 degrees – which is the most common level of accuracy. Class A is +/- 2 degrees and class C +/- 10. The pilot may also request steering information using Q-code terminology, for example, QDM is magnetic heading to the VDF station.

Emergency Locator Transmitter (ELT)

Another useful tool a pilot in an emergency can utilize is the emergency locator transmitter (ELT). The ELT is a self -contained, self-powered radio transmitter, designed to transmit a signal of distress. A pilot would not normally activate the ELT during a flight, but rather the ELT is an important communication device in a post-crash scenario. Most aircraft are required to be fitted with an ELT, although there are a few exceptions. In most cases, it is located in the tail of the aircraft.

The ELT is generally has two modes; armed or on. When the ELT is switched to 'armed' mode, it is activated by a g-switch, which is triggered should the aircraft experience a certain level of deceleration (e.g. crash landing). When the ELT is switched 'ON', it begins to transmit a signal straight away. When active, most ELTs will transmit on three frequencies, 121. 5 MHz, 243 MHz (military), and 406 MHz (UHF). The first frequency we have already touched on in this chapter. The ELT will make a very distinct sound on 121.5 MHz, potentially alerting a nearby air traffic control unit or any passing aircraft monitoring the frequency. The third frequency (406 MHz) is not a frequency that can be tuned up on a standard aircraft radio unit, but rather it is detected by the search and rescue satellite network that can determine the ELTs location.

As a pilot in a non-emergency situation, it is good practice to listen on 121.5 MHz at the end of a flight. This is for two reasons. First, if another aircraft ELT is activated nearby, you may be able to hear it and alert the nearest air traffic service. Second, if you have landed firmly, your ELT g-switch may have triggered. If you have inadvertently activated your ELT, turn it off and notify the nearest air traffic service.

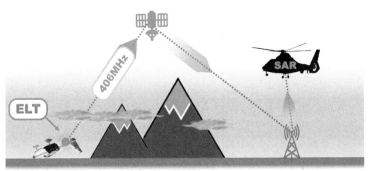

Figure 12.1: *An ELT transmitting on 406 MHz can transmit a signal to satellites which can be used by search and rescue (SAR) to locate a stricken aircraft.*

The ELT is far from perfect. ELTs transmitting on 121.5 MHz have a limited range, which can easily go unnoticed if there are no nearby receivers (e.g. aircraft or Tower). Furthermore, a search and rescue operation using 121.5 MHz will find it difficult to locate the exact crash site. During a crash, an ELT may fail to activate or become disconnected from its aerial. For this reason, after a crash landing, it is advised to check the ELT is active (if it is safe to do so) and the aerial is still connected.

For all flights, it is essential to inform someone about the flight, including the intended route and destination. Ideally, this would be in the form of a formal flight plan with a search and rescue time stipulated. If the flight fails to arrive at its intended destination, the relevant authorities would then have the necessary information to find the aircraft and occupants swiftly. Unlike a car crash where help may be in the next passing car, a pilot might not see another person for days.

> One of the longest post-crash survivals is of Bob Gauchie. Bob was flying solo across northern Canada in 1967 in stormy conditions. His aircraft ran out of fuel, forcing him to make a forced landing in a remote area, isolated from the outside world. Bob endured 58 days in freezing conditions before being rescued.

By utilizing all the benefits of an air traffic control service, you can greatly increase your chance of a positive outcome during an emergency. Before a flight, ensure a flight plan is filed, seek assistance early, and don't be afraid to declare an emergency.

CHAPTER 13: **RADIO EQUIPMENT**

PILOT: Approach, what's the tower?
APPROACH: Well, it's a big tall building with windows or 126.9, which do you want to know?

The final part of our flight radio journey takes a look at the typical radio equipment found in an aircraft and emerging technologies in this area. It is not possible to cover all types of radio equipment, but most have some common features. Firstly, the names used for the flight radio can vary; from COMMs, Com box, audio box, or VHF radio unit. It is also common to have the VFR navigation receiver combined with the flight radio (to select a VOR or ILS frequency). Before tuning up the radio unit, you first need to find the correct frequency. Frequencies can be found in a range of locations, with the aerodrome chart a common source (or a similar publication), as shown in Figure 13.1. During a flight, the controller will normally provide the next frequency you require (e.g. *contact Hamilton Tower on 118.3*).

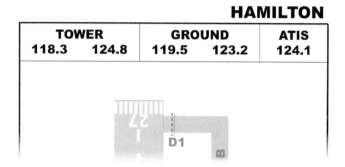

Figure 13.1: *An aerodrome chart is one source of information of frequencies required for a flight.*

The radio unit in most aircraft is called a transceiver, which means it is both a transmitter and receiver (able to send and detect radio waves). The complexity of radio equipment varies, with some old VHF radio units only able to select a single frequency. Modern radio units allow the selection of multiple frequencies (but pilots can only talk on one at a time), with active and standby frequencies displayed, as shown in Figure 13.2. The standby frequency box displays the frequency that can be changed, which can then be moved over to the active frequency. When possible, try and anticipate what frequency you will require next and have it tuned up in the standby frequency box. For example, if you are currently on the Ground frequency, Tower is likely the next person you will speak to, therefore you can have the Tower frequency all ready to go. By tuning this frequency up early during a low workload phase, you reduce the risk of selecting the wrong frequency later on.

Figure 13.2: *A typical VHF radio, with multiple frequencies (VHF1, VHF2, VHF3). The knob on the right allows the 'Standby' frequency to be changed, which can be switched over to 'active' by depressing the button in the middle.*

It is common for modern radio units to allow multiple frequencies to be selected. But why so many, when you can only talk on one frequency at a time? The extra frequencies are helpful if you need to monitor a frequency, such as an ATIS. For example, if you are in controlled airspace, tracking towards your destination, you will want to obtain the ATIS in good time. With the aid of another selectable frequency, you can still remain active on your controlled frequency (e.g. Approach), and listen to the ATIS on the second frequency.

In terms of general use, ensure your microphone/ headset is plugged in correctly before operating the radio. Make sure the electrical switch is turned on (some aircraft may have a separate avionics power switch). Ensure the volume is set to an appropriate level (this can be tested by monitoring the ATIS). Some radios may also have a squelch knob. This control acts as a filter to reduce static (interference) on the radio.

The microphone component of the radio can vary and generally falls into two categories; boom microphone or handheld microphone. The handheld microphone, as the name suggests is held in your hand, with the microphone switch located on the handheld device. Most aircraft are now equipped with a boom microphone, which means you can operate the radio without taking your hands off the control column. The boom microphone has the microphone attached to your headset, with the microphone activated with a switch on the control column. It is important to keep the microphone at a constant distance from your lips, but not too close (as a general rule, your lips should not touch the microphone). As previously discussed, make sure the microphone button is released once you have finished transmitting.

VHF flight radio is the mainstream of communicating with air traffic control, but there are a few other systems; old and new. When flying in areas with limited VHF reception, a separate radio may be required for Long Range HF (High Frequency) communication. As the name suggests, this radio is used for long-range communication, but as discussed in chapter two, HF transmissions can suffer from poor quality. The HF radio is a separate unit but works in a similar way to a VHF radio unit. The main visual difference is that the frequencies used are considerably different than VHF. HF frequencies will sit between 3 MHz and 30 MHz. For example, an HF frequency may be displayed as 3.485 or 10.051 MHz.

Technology has provided a solution to the limitations of HF radio in the form of Controller Pilot Data Link Communications (CPDLC). CPDLC is more likely to be found in an airliner crossing an ocean but may be used in smaller aircraft operating in remote areas. In its basic form, the system can send short, text-based messages between an aircraft and the controller. For example, CPDLC may be used for pre-departure clearances, altimeter settings, transfer of communication, initial contact, and emergency messages. Aircraft require the appropriate equipment to use CPDLC, and voice communication still remains the primary means for pilots and controllers to communicate.

ADS-B is also providing some potential improvements to the flow of information pilots receive. As discussed in chapter eight, the main mode of ADS-B is known as ADS-B Out, whereby the unit sends the aircraft's position to a ground station. However, units with the ADS-B-In function are able to receive information (but may not be available in all areas). This may include the position of other aircraft or weather information. ADS-B In can greatly enhance a pilot's situational awareness, with information that could previously only be provided by an air traffic control unit.

When you jump into a new aircraft, you can encounter a considerable range of radio equipment. It is important for pilots to become familiar with the unique radio equipment fitted to their aircraft. Spend some time identifying the radio functions and controls. The last thing you want to be doing is giving yourself a crash course on how your radio operates while flying the aircraft.

CONCLUSION

The aircraft radio is an amazing piece of equipment. By pressing down on the microphone switch you can have a conversation with a wide range of people; from the local controller to other pilots in the area. But operating the radio also comes with responsibility. To avoid the serious consequences that can result from miscommunication, it is essential that all pilots adopt the common radio practices.

We started our journey exploring how the radio works. By exploring the basic components of a radio wave, you can now see aviation communication is squeezed into a tiny section of the radio spectrum. We saw that the building blocks of flight radio are based on how numbers and letters are pronounced, and common phrases and words. We then put these building blocks into practice and took a flight. We saw the range of radio calls you might make, including the taxi and take-off calls at a busy aerodrome, with a Ground or Tower operating. But just as important are the radio calls at non-towered aerodromes. Radio calls are essential in uncontrolled airspace as they help keep everyone else aware of each other's position. The final chapters demonstrated how useful the radio can be during an emergency. With a push of a button, you can receive assistance from a range of sources to help bring a stricken aircraft home safely.

Aviation is fast evolving, with new technologies making their way into the cockpit. Flight radio is no exception, with a range of new technologies helping to overcome some of the current limitations of VHF communication. However, the primary method of communicating will remain the flight radio for some time. Whether you are a pilot who is just experiencing the joy of flying for the first time or flying has been your passion for countless years, it is essential to apply the common communication principles to keep the skies safe.

INDEX

ADS-B, 69

Aircraft call sign, 33

Aircraft radio unit, 121

Broadcast, 54

Channel spacing, 14

Clearance, 46

Conditional clearances, 47

Departure radio calls, 63

Distress call (Mayday), 112

Emergency Locator Transmitter (ELT), 117

Frequency, 7, 23,

Frequency transmissions, 51

Flight following, 73

IFR radio , 94

Initial call, 57

Light signals, 106

Non-towered (uncontrolled) radio procedures, 64, 91

Radar, 66

Radio failure, 101

Radio spectrum, 8

Radio waves, 4

Read-back, 48

Speechless techniques, 108

Squawk, 71

Station call sign, 37

Take-off radio calls

Taxi radio calls, 59

TCAS, 79

Test procedure, 53

Time, 23

Transmission of letters, 18

Transmission of numbers, 20

Transponder, 67

Traffic pattern (circuit), 85

Urgency call (Pan Pan), 114

Very high frequency, 11

VFR flight plan, 75

Wavelength, 6

BOOKS IN THIS SERIES

Book 1: Human Factors for the Private Pilot

From the dangers a pilot faces when straying too high in an oxygen-deprived atmosphere to the way the brain attempts to process the enormous amount of information obtained during a flight. This book is for pilots and non-pilots to explore the vast number of factors that can influence a pilot's ability to fly an aircraft safely.

Book 2: Aviation Weather for the Private Pilot

Aviation weather is a wondrous and frightening subject. Pilots can encounter a range of weather conditions on just a single flight, from a towering thunderstorm that can toss an aircraft around like it is in a washing machine, to thick fog in which pilots will struggle to see just a few feet in front of the aircraft.

Book 4: Principles of Flight

Principles of flight is one of the fundamental topics a pilot must master to operate an aircraft safely. Pilots have so much control over the various forces acting on the aircraft, but flying an aircraft also comes with responsibility. To avoid flying an aircraft beyond its limits, pilots must respect the principles of flight. Turning too sharply, flying too slowly, or overloading an aircraft can all have dire consequences.

Book 5: Flight Navigation

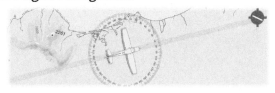

A pilot can travel a considerable distance, across a range of different landscapes on a single flight; from rugged bush, oceans, mountainous terrain and deserts. No other mode of transport offers such freedom. To arrive safely at a distant destination, pilots must understand the key components of *Flight Navigation*. Flying the wrong heading, underestimating the time and fuel for the flight can all have serious consequences.

Printed in Great Britain
by Amazon

76714555R00081